Y0-BRI-784

COMMUNICATION, STORAGE AND RETRIEVAL OF CHEMICAL INFORMATION

ELLIS HORWOOD SERIES IN INFORMATION SCIENCE

COMMUNICATION, STORAGE AND RETRIEVAL OF CHEMICAL INFORMATION
Editors: J. E. ASH, P. A. CHUBB, S. E. WARD, S. M. WELFORD, P. WILLETT

CHEMICAL INFORMATION SYSTEMS
J. ASH, Information Services Consultant and E. HYDE, ICI Pharmaceuticals Division

CHEMICAL NOMENCLATURE USAGE
R. LEES and A. SMITH, Laboratory of the Government Chemist, London

DESIGN, CONSTRUCTION AND REFURBISHMENT OF LABORATORIES
Editors: R. LEES and A. SMITH, Laboratory of the Government Chemist, London

COMMUNICATION, STORAGE AND RETRIEVAL OF CHEMICAL INFORMATION

JANET E. ASH, B.Sc., M.Sc.
Consultant (Information Services)
PAMELA A. CHUBB, B.Sc.
Consultant, Logica UK Limited
SANDRA E. WARD, B.Sc.Hons., Ph.D.
Head of Information Services, Glaxo Group Research Limited
STEPHEN M. WELFORD, B.Sc., M.Sc., Ph.D.
Research Assistant, University of Sheffield
PETER WILLETT, M.A., M.Sc., Ph.D.
Lecturer in Information Studies, University of Sheffield

ELLIS HORWOOD LIMITED
Publishers · Chichester
Halsted Press: a division of
JOHN WILEY & SONS
Chichester · New York · Ontario · Brisbane

First published in 1985 by
ELLIS HORWOOD LIMITED
Market Cross House, Cooper Street, Chichester, West Sussex, PO19 1EB, England

The publisher's colophon is reproduced from James Gillison's drawing of the ancient Market Cross, Chichester.

Distributors:

Australia, New Zealand, South-east Asia:
Jacaranda-Wiley Ltd., Jacaranda Press,
JOHN WILEY & SONS INC.,
G.P.O. Box 859, Brisbane, Queensland 40001, Australia

Canada:
JOHN WILEY & SONS CANADA LIMITED
22 Worcester Road, Rexdale, Ontario, Canada.

Europe, Africa:
JOHN WILEY & SONS LIMITED
Baffins Lane, Chichester, West Sussex, England.

North and South America and the rest of the world:
Halsted Press: a division of
JOHN WILEY & SONS
605 Third Avenue, New York, N.Y. 10016, U.S.A.

British Library Cataloguing in Publication Data
Communication, storage and retrieval of chemical information. –
(Ellis Horwood series in chemical science)
1. Chemistry – Information services 2. Chemistry – Bibliography
I. Ash, Janet E.
540'.7 QD8.3

Library of Congress Card No. 84-25170

ISBN 0-85312-571-6 (Ellis Horwood Limited)
ISBN 0-470-20145-2 (Halsted Press)

Typeset by Ellis Horwood Limited.
Printed in Great Britain by The Camelot Press, Southampton.

Table of Contents

Preface

The Chemical Structure Association (CSA) Seminar 'The Future of Chemical Documentation' at the University of Exeter in September 1982, from which this book has been derived, had two major objectives: (1) to bring together all those who need or work with chemical information to discuss present systems and to predict and plan future developments; (2) to ensure that the discussions involved those who provide access to published information as well as those who are responsible for the development of internal data handling systems in large companies.

The second objective was stimulated by an awareness that over many years, techniques for handling the two types of information had diverged increasingly; within industry, priority had been given to the development of the databank to store a range of physico-chemical and biological data, accesssible by a variety of sophisticated retrieval techniques based on the chemical structure. In the published literature field, developments in computerized information handling had, until the recent introduction of structure-based searching systems such as CAS ONLINE and DARC, centred on the development of text-based searching techniques for accessing bibliographic information. Because these two development paths were beginning to converge it seemed likely that the two groups would benefit from exposure to each other's activities.

Another reason for bringing these groups together at this time was the enormous growth in information technology which has already taken place and which will certainly continue. Information technology provides not only the background against which future information systems will evolve but also the stimulus for much of the evolution. Since it can also offer the opportunity to develop a single interface for the user to access both his own local data collection and the world's published literature, it was both logical and necessary to use the expertise of 'internal' and 'external' information specialists to consider the possibilities.

Any gathering held to discuss chemical information is sure to attract large numbers of information scientists. In addition, the importance of two further groups to any real discussion was recognized – the database vendor and the chemist. The former needed to be involved because, in general, vendors have not appeared to be particularly susceptible to comments on their existing services and suggestions for improvement, nor do they seem to survey user needs actively. Also, vendors are slow to adopt the results of information research. The chemist is notoriously difficult to attract to discussions on information services yet as the end user, the chemist should be the prime focus of any discussion on chemical information requirements. This is particularly important at a time when new technology is laying the foundations on which major changes can be constructed.

This book is therefore directed towards those who generate and those who use chemical information, towards those involved in its organization and distribution. Covering the three distinct areas of chemical information handling – bibliographic, numeric, and structural – it aims to review fully all current techniques as well as the latest developments in research and relates them to user requirements. The authors have based their chapters in part on the material presented at the conference and on the contributions of those who participated in formal and informal discussion. The conference material has been heavily augmented to ensure that the information presented here gives a more thorough overview of the selected topics. The authors are aware, however, that the book's content is not absolutely comprehensive and that information, while current at the time of going to press, dates extremely quickly in such a fast-moving area.

The book begins by considering the information needs of chemists, looking at their working environment and commenting on the limitations of existing techniques and services. Following this, the book takes an historical perspective beginning with the longer established methods of handling textual information in Chapter 2. This reviews both primary and secondary publication and also includes sections on patents and the major printed reference books. Chapter 3 describes the computerization of bibliographic information pioneered by the growing online industry and includes references to the development of the electronic journal. The growing interest in the availability of databanks which contain directly usable numeric or textual information is reflected in Chapter 4 which describes the major databanks so far developed and problems and factors affecting the development of databanks.

The most important aspects of chemical information handling are obviously those techniques developed for storage and retrieval of chemical structures. They are extensively examined in Chapters 5 to 9 which cover existing methods for the representation and registration of structures in computer systems, the techniques developed for substructure searching, and software for the storage and retrieval of chemical structures, both commercially available packages and those developed for use with particular databases. The various approaches so far taken in attempts to develop generally applicable reaction indexing systems which are of prime interest to the chemist are described. Techniques used to exploit stored chemical structural information are also considered and include structure–activity methods, molecular modelling and computer-aided chemical synthesis.

Since all of the topics covered in these chapters feature computerized information handling techniques, Chapter 10 examines current developments in computer and telecommunications technology. The general principles of graphics hardware and software are also discussed.

Chapter 11 highlights the important themes identified in previous chapters, indicating the main growth points for future developments and highlighting possible problems. But reviewing the most significant areas of change proves enormously difficult when so many influential developments will come undoubtedly from outside the traditional information world.

Work on individual chapters was undertaken as follows:

J. E. Ash – Chapters 2 and 3
P. A. Chubb – Chapters 4 and 11
S. E. Ward – Preface and Chapter 1
S. M. Welford – Chapters 5, 6, 7 and 9
P. Willett – Chapters 8, 9, 10 and 11

Particular acknowledgement is made to Dr Peter Murray-Rust of Glaxo Group Research Ltd who made a considerable contribution to the sections on computer graphics and Dr Wendy Warr of ICI plc Pharmaceuticals Division on whose presentation Chapter 7 is based. Also to Dr Paul Rhyner of Ciba Geigy plc, many of whose ideas on the future development of chemical information systems are included in Chapter 1. Mr Ernie Hyde of Fraser Williams (Scientific Systems) Ltd and Dr David Bawden of Pfizer Central Research Ltd are also thanked for their helpful comments. The conference was organized by Jane Whittall (then Gaworska) of Beecham Pharmaceuticals Ltd and her considerable contribution to the conference programme must be acknowledged here.

Finally, we should like to thank all the following contributors to the conference for permission to use material from their lectures in the preparation of this book.

Mr L. S. Adler, Chemical Industries Association
Notification of Chemical Data and Questions of Confidentiality

Dr F. H. Allen, Crystallographic Data Centre, Cambridge
Molecular Structures Rejuvenated – the Role and Utilization of the Cambridge Structural Database

Mrs F. H. Barker, Royal Society of Chemistry, Nottingham
Development of Chemical Databanks

Dr K. P. Barr, The British Library Lending Division
Obtaining Chemical Information

Dr J.-C. Bonnet, Télésystèmes-Questel
Chemical Databases

Professor R. T. Bottle, The City University
Present Methods of Communicating Chemical Knowledge

Dr J. Brandt, Technical University of Munich
A Systematic Classification of Reactions by Electron Shift Patterns

Dr H. D. Brown, Merck, Sharpe & Dohme, USA
Information Needs of Medicinal Chemists in the Pharmaceutical Industry

Dr J. Buckingham, Associated Book Publishers Ltd
Heilbron's Dictionary of Organic Compounds DOC 5

Mr P. T. Bysouth, Glaxo Group Research Ltd, Greenford
Retrieval of Substance Information from the Biochemical Literature

Dr R. Coleman, The Government Chemist
Introductory Speech

Dr C. Cundy, ICI plc, Runcorn
A Personal View of the Generation and Use of Information in Inorganic Chemistry Research

Miss J. Dalton, Beecham Pharmaceuticals Ltd
Compound Registration

Dr J. B. Davis, Health and Safety Executive
The New Substances Notification Scheme – Scientific and Operational Aspects

Mr R. Dean, Excerpta Medica, Amsterdam
DIOL – Drug Information Online

Dr A. Everett, Wellcome Research Labs., Beckenham, Kent
The Information Needs of Physical Chemists

Dr E. Garfield, ISI
Chemical Information Processing at ISI

Dr M. Hann, G. D. Searle
How does a Bench Chemist's Work Benefit from New Substructure Technology?

Mr. A. Haywood, Exxon Office Systems
Trends in Telecommunications, Communications and Hardware

Dr S. R. Heller, Environmental Protection Agency, Washington DC
Experience in the Development of CIS

Dr P. J. Hills, University of Leicester
Communicating at Conferences: The Presentation of Chemical Papers

Dr P. L. Holmes, Blackwell, Technical Services, London
Document Delivery Developments

Dr R. Hyde, Wellcome Research Laboratories, Beckenham
Data Requirements for QSAR Studies

Dr R. A. Y. Jones, University of East Anglia
Naming Organic Transformations

Dr H. Kaindl, Sandoz Ltd, Basle, Switzerland
The Function of the Internal Database – Burial Ground or Intelligence Service?

Dr A. Kolb, IDC, Frankfurt, West Germany
Are Fragmentation Codes Obsolete?

Dr R. Langridge, University of California, USA
The Opportunities for Graphics Techniques in Chemistry

Dr R. Linford, Leicester Polytechnic
Information Needs – the Academic Viewpoint

Dr J. S. Littler, University of Bristol
The Nomenclature of Transformations and Mechanisms – A View from IUPAC

Dr R. Luckenbach, Beilstein Institute, Germany
The Beilstein Handbook – After the First Centennial

Dr A. Mackay, Birbeck College, University of London
Information and Dimensionality – the Communication of Chemical Knowledge

Dr D. S. Magrill, Fisons Pharmaceuticals Ltd, Loughborough
A Registry Number for Today

Dr S. Marson, Molecular Design Ltd
Interactive Graphics Systems for Chemical Research

Dr G. Moreau, Roussel-Uclaf
Substructure Search System

Mr A. Negus, Consultant
Software Trends

Mr P. Nichols, Pergamon InfoLine Ltd
Specialized Chemical Files from InfoLine

Mr P. Norton, Derwent Publications Ltd
Coding and Retrieval of Markush Structures in the Derwent Central Patents Index – Past, Present and Future

Miss M. O'Hare, The British Library, R&D Division
Translating Research Results in Chemical Information into Practice

Dr E. Onerato, ESA/IRS
The Chemical Abstracts Service Bibliographic File on ESA/IRS

Dr D. P. J. Pearson, ICI Plant Protection Ltd
Problems Associated with the Use of Computers as Aids to Chemical Synthesis

Mr J. Revill, Beecham Pharmaceuticals Ltd
European Inventory of Existing Chemical Substances EINECS

Mr J. F. B. Rowland, The Royal Society of Chemistry
Do we need the Scientific Paper?

Miss K. Shenton, SDC Search Service
You and ORBIT: the Chemist's Right

Dr J. Sibley, Shell International Petroleum, London
Patents, the Undervalued Resource

Mr B. Stanford-Smith, National Computing Centre Ltd, Manchester
Technology and the Availability of Information

Dr H. W. D. Stubbs, The Royal Society of Chemistry, Nottingham
The Future of Chemical Documentation – Economic Aspects

Professor R. L. M. Synge, University of East Anglia
Some Experience with the Secondary and Tertiary Services

Dr S. Terrant, American Chemical Society
Online Access to Full Text ACS Primary Journals

Mr S. Vogt, Lockheed Dialog
The Future of Chemical Information on Dialog

Dr W. A. Warr, ICI Pharmaceuticals Ltd
Software

Dr S. M. Welford, University of Sheffield
Towards Simplified Access to Chemical Structure Information in the Patent Literature

Dr P. Willett, University of Sheffield
Some Automatic Approaches to the Indexing of Chemical Reactions

Dr P. W. Williams, Userlink Systems Ltd
Improving Online Access by using Microtechnology

Professor W. T. Wipke, University of California
Reaction Storage and Retrieval

S. E. WARD
Ware
May 1984

1

The information needs of chemists

1.1 INTRODUCTION

Chemistry, since its liberation from the secretive practices of the alchemists, has always been that branch of human endeavour where information is most effectively communicated and best organized for retrieval. This is partly because records of experiments in chemistry have lasting practical value – it is not uncommon to find a chemist following preparative details which were published more than a century ago or confirming the appearance of a substance with a physical description of similar antiquity. Also, chemical substances, by their nature, lend themselves to systematic documentation and indexing, based on concepts of chemical structure which have remained constant over a relatively long period and which are largely independent of language. Most of all, it is because the chemical task has bred a co-operative spirit among chemists which has fostered the exchange of information and encouraged the evolution of formal documentation procedures.

Development of systematic information handling in chemistry is very firmly rooted in the efforts of chemists themselves. The chemist elucidated the principles of chemical structure, and, following this, invented nomenclature systems and other techniques for describing structures. The chemist, largely through the various national professional chemical societies, has been and continues to be responsible for the development and production of most chemical literature. Cooperation between groups of chemists has also led to the development of certain key data collections.

Although the chemist still produces chemical information, in recent years he has become divorced from overall control of its communication and retrieval. The two major factors contributing to this separation are, firstly, the huge increase in the volume of the chemical literature and, secondly, the growth in the use of computers for information handling. Over the past forty years the annual volume of chemical papers has increased roughly five-fold, while novel experimental techniques have greatly increased the information content of these documents. As a result, at least in industry, much of the chemist's role in information gathering and analysis has been seconded to the information scientist for convenience and time saving. The centralization of searching expertise, initially required for the most efficient exploitation of online databases, has persisted in many organizations. Thus many chemists are still second-class information citizens making do with the traditional methods of searching the chemical literature, and frequently only the librarians and information scientists have the full spectrum of modern techniques at their disposal. The development of improved access methods to online chemical databases is reversing this trend although chemists are only slowly adopting online techniques.

Not only has the chemist become separated from the published literature but, in industry, this separation extends to the chemist's own research results. The development during the 1960s and 1970s of centralized industrial databanks maintained and accessed by information scientists is only now being succeeded by the introduction of online systems which are searchable by chemists as part of their daily routine and to which chemists themselves contribute data directly.

As the rest of this book will show, chemical information techniques are evolving extremely rapidly. The systems and services available to the chemist are already among the most sophisticated of any, yet there are still a large number of outstanding problems and opportunities. It is therefore appropriate to focus on the information needs of the chemist in this opening chapter.

There has been little or no formal documentation of the chemist's requirements for information. A complete survey of the enormous range of chemical activity and the variety of working environments is beyond the scope of this book. Likewise the considerable need of the non-chemist and non-scientist for chemical information cannot be considered. Instead, two types of chemists are selected for review, the teacher, and the research worker, particularly the research and development chemist in the pharmaceutical industry, an industry which has long recognized the importance of effective information provision for research and development. Both groups have exerted substantial influence on the evolution of chemical information services and the role and scientific information needs of both are examined. Recommendations are made for improvements in chemical information handling based on limitations noted in existing systems and services. Requirements for training chemists in information retrieval techniques are also included since the lack of training is recognized as being one of the causes of lack of acceptance of modern information methods by the chemist [1]. For completeness the changing role of the chemical information scientist is also considered since the information scientist will continue to play a crucial part in chemical information for some considerable time.

1.2 THE RANGE OF INFORMATION REQUIRED BY CHEMISTS

The data and text generated by chemists for use by chemists comprises both generally available published material and information generated within a particular organization for its own purposes.

The chemical information requirement of the chemist stretches beyond chemistry itself. Chemistry is an enabling science: chemical substances are ubiquitous and many other scientific disciplines study the effect and behaviour of chemicals. Thus, in addition to requiring chemical information, the chemist requires and needs to understand information from other sciences to support his own chemical activity. The chemist also needs to present his own information to other disciplines in a comprehensible way.

The chemist therefore is highly unlikely to be able to satisfy the need for published information from a few 'pure' chemistry journals. Neither, in industry, will the chemist restrict his internal information interest to data generated within the chemical department. The breadth of disciplines which can be seen to interact with chemistry is demonstrated in Fig. 1.1.

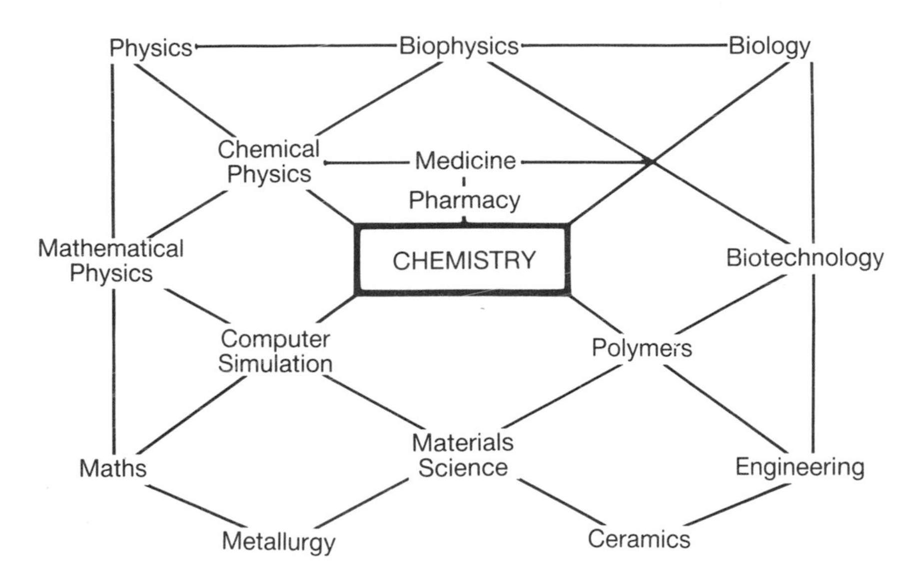

Fig. 1.1 – The breadth of modern chemistry.

The techniques and services for handling of information in these related disciplines is therefore equally interesting to the chemist as the handling of chemical information.

1.3 THE NEEDS OF RESEARCH AND DEVELOPMENT

1.3.1 The chemist in pharmaceutical research and development

The pharmaceutical industry supports a considerable chemical research and development activity. It has therefore invested heavily in information resources, having large information departments, significant budgets for the purchase of published information, and sophisticated systems for the capture and retrieval of internally produced information. Chemists working in pharmaceutical R&D thus provide a model of what can be achieved in an information conscious and information-rich environment.

Table 1.1 shows the various stages in a pharmaceutical research project for drug discovery and development and the interrelationships of the required scientific disciplines. Of key importance is the role of the chemist – the medicinal chemist, the physical chemist concerned with analytical techniques, and the development chemist.

The medicinal chemist wishes to synthesize new chemical entities with novel or improved biological activity. This frequently requires the synthesis of a wide variety of chemical classes. Even when plant biochemicals or microbial broths are the starting point for a new product candidate, synthetic modification may be required to improve the physico-chemical properties, to increase biological activity and potency or to strengthen patent protection. The medicinal chemist must interact with all the scientific disciplines available in his organization.

The physical and analytical chemists within the industry exploit many techniques to support the development of new medicines – ultraviolet (UV), infra-red (IR) and nuclear magnetic resonance (NMR) spectroscopies, mass spectrometry (MS), chromatography, electrochemistry, X-ray analysis, and methods for determining general properties such as pK_a, dipole moments, and partition coefficients. Work will include the confirmation of the structure of new entities, the development of analytical techniques to determine drug purity, and the support of groups seeking to determine mechanisms of action and metabolic profiles.

The initial synthesis and evaluation of a compound requires usually only small amounts of material. When the decision is taken to develop that compound as a drug candidate much larger quantities will be required for pharmaceutical formulation, for testing in animals and man, and for the development of analytical methodology. The responsibility for taking a laboratory method for the synthesis of a drug and scaling it up or developing a new route for large-scale synthesis lies with the development chemist. If this work is to be successful, a safe, reliable, and cost-effective manufacturing process must be developed to provide material of high purity defined to an analytical specification. Often the synthesis used in research will not be appropriate for scaling up and a new method of preparation must be sought. Early process investigation is started on a small scale and is concerned with using cheaper starting materials, optimizing yields and throughput, and reducing the number of stages. The process is then scaled up into large glassware and into general-purpose pilot plant. Finally, a manufacturing process is commissioned.

Table 1.1 – Stages in a pharmaceutical research project.

Disciplines	Primary research: drug discovery	Product development	Clinical trials and pre-product launch	Marketing of drug
Biological research	Fundamental research to define biological target Development of screening methods Primary screening Secondary screening Acute toxicology	Screening of follow-up compounds Additional biological studies Drug interaction studies		
Chemical research	Preparation of compounds Filing patents	Preparation of follow-up compounds and patent examples		
Chemical development		Synthesis development Preparation of drug for toxicology, pharmacology and clinical studies Preparation of radiolabelled drug	Transfer of development methods to production	Chemical manufacture
Biochemical pharmacology		Development of assays for drug in biological samples Metabolic studies in animals Assay of samples from toxicology and human volunteer studies Metabolic studies in man	Assay of samples from patients	Assay of samples from patients
Pharmacy		Formulation development Preparation of formulated drug for toxicology and clinical studies Stability trials on raw and formulated drug Development of assays on raw and formulated drug Analysis of raw and formulated drug	Transfer of formulation and analytical methods to production	Development and stability of new dosage forms
Physical chemistry	Determine identity and purity	Develop assays; analyse raw and formulated drug		
Toxicology and histopathology		Subacute toxicology Long-term toxicology Reproductive toxicology Carcinogenicity studies		
Clinical		Human pharmacology Preparation of clinical trial protocols	Clinical trials	New claims Additional clinical trials for product support once marketed

1.3.2 Information requirements of the pharmaceutical chemist

The fundamental knowledge required by a pharmaceutical chemist will include knowledge of extant therapeutic agents, insight into the complex facets of drug design, understanding of classical and contemporary synthesis, and up-to-date knowledge of the principles of organic chemistry, familiarity with patent systems and an ability to develop and exploit techniques for the analysis and evaluation of data. A variety of other information needs complement this basic knowledge.

Both the medicinal and development chemist will need to maintain current knowledge of new and established reactions and of new compound types. The development chemist will be concerned with one key structure. He will thus wish to be aware of established routes to the target molecule or routes to related molecules which might be adapted. Information on the likely mechanism of reactions and factors influencing them will be required, as will be knowledge of distinctive properties of the target molecule which may allow the progress of the reaction and its yield to be simply measured. Knowledge of commercial availability of compounds in research and bulk quantities is also required.

Hundreds and thousands of specific compounds are likely to be prepared and evaluated in the course of launching one new medicinal product. This necessitates access to a large volume of data held both in the published literature and in proprietary files. The medicinal chemist will wish to establish competitor interests and to compare and interpret data on his own compounds with other in-house and external data. Information from disciplines other than chemistry (biology, biochemistry, toxicology, pharmacy and medicine) will be required if the medicinal chemist is to develop compounds which are capable of exerting the desired effect *in vivo* without any undesirable activity. Up-to-date knowledge of research in other sciences is thus required. Whilst a chemist can cope with quantitative measurements and qualitative information within his own speciality the provision of evaluated data expressed in a standard, unambiguous and easily interpreted form is essential for effective communication from other disciplines.

The manipulation of biological, structural and physico-chemical data to determine trends and discover leads is a continually growing need of the medicinal chemist. This will be done both by intuition or serendipity and by utilizing in-house and commercial software to manipulate data generated internally and that extracted from the literature. Physico-chemical data will therefore be required not only for confirmation of structure but for use in structure–activity work and, if not available, will need to be measured or derived from existing data.

In addition to maintaining awareness of new techniques and new instrumentation, a high proportion of the physical chemist's information need is also concerned with data analysis. The volume of data generated internally is large. A total generation volume of several hundred thousand measurements per annum is not uncommon. The physical chemist is thus faced with the problem common to all chemists of effectively documenting his own information and of relating and comparing this to data, several times greater in magnitude, existing externally, in order to determine the identity of known and unknown compounds. As for other fields only a proportion of this data is published. Much sits in the internal databanks of other companies; even more exists within other non-industrial laboratories. Little is available in a standard format for simple interpretation.

Techniques such as mass spectrometry are highly dependent on the availability of large reference databases such as the Mass Spectrometry Data Centre (MSDC) database. Identification of totally novel structures can be aided by heuristic programs which enable derivation of structure by comparing data in the unknown molecule with data from known structures. These techniques currently require large programs, large databases, and significant computing power but are already seen within the pharmaceutical industry where they offer the opportunity to lower the level of spectroscopic interpretive skill, thus releasing intellectual effort for the key objective, that of developing new medicines.

The pharmaceutical industry's chemists therefore require access to a wide range of information including full textual information and data for browsing and consideration and also for immediate use. The information exists both within the company and externally and will be obtained from colleagues, via information scientists, or directly from published sources. External and internal information is not simply restricted to purely chemical information but may originate from a range of related disciplines. Increasingly, use of information includes both absorption into the chemist's own knowledge base and the computer assisted manipulation of data.

The chemist requires not only good access to current information of interest but efficient retrospective access to selected material. A variety of access points are required to ensure retrieval tools can satisfy all known and unanticipated needs. The key access point is, and will continue to be, the chemical structure. Chemists in this respect have two major requirements [2]:

(a) to identify the existence of and to retrieve information on a particular structure – 'novelty searching';
(b) to identify groups of compounds each containing a structural moiety of interest – 'substructure searching'; the substructure may be of interest because it is a synthon or part of a synthon (a functional group held in a masked form to simplify a synthetic reaction) or because it is thought to be important in conferring a particular type of activity on the substance, e.g. a pharmacophore.

A third key access point required will be the data itself and the chemist will wish to identify easily whether a particular data type is available for a compound and also to find molecules with data values within particular ranges of interest.

In addition to obtaining information generated by others the chemist will also be required to communicate and document information. The company will have well developed standards for laboratory documentation, and progress of work achieved will need to be regularly reported to colleagues and superiors. The chemist may also be involved in the compilation of patents and will certainly wish to achieve publication in the general chemical literature, where research of significance has been achieved and the company's interests will not be prejudiced by so doing.

One final point should be made in this section and this concerns simplicity. The chemist can afford to allocate only a limited proportion of effort to retrieving information. His needs for acquiring and communicating information must be met in the minimum time. Simple tools are needed for immediate access to information required for immediate use. Information on current topics of interest must be able to be acquired with little effort and searching of internal and published information must be similarly simple and efficient. Dissemination of his own research work too must be as non-laborious as possible.

1.3.3 Internal information services used by and developed for the pharmaceutical chemist

Published information in its various forms – the primary, secondary and tertiary literature including patents, indexes to reactions, and the development of databanks are well described in subsequent chapters and these should not be preempted. This section therefore concentrates on the internal information systems developed within the pharmaceutical industry. A product of the chemist's own needs coupled with those of scientists in other disciplines, these systems provide chemists with an integrated route to retrieval of information from a variety of sources. In this sense, these systems are superior to the present compartmentalized external services.

Without exception all pharmaceutical companies have developed complex and now largely computerized internal information systems [3, 4]. A simplified picture of the components of such systems is given in Fig. 1.2. Central to the internal information system of any pharmaceutical company is the company registry, a file of structural information on key compounds synthesized as intermediates or for biological examination. Normally it is compulsory for a chemist to notify new substances to this registry once their structure has been confirmed. The number of compounds held in such files is high. Merck, Sharp & Dohme's registry system contains 140,000 compounds, that of ICI Pharmaceuticals some 200,000 compounds, that of Glaxo Group Research some 60,000 compounds. Methods of structural representation used have included fragment codes, molecular formula, and Wiswesser Line Notation. Each structural record is associated with the company registry number. Within the pharmaceutical industry the hierarchical classification of compounds into parent (a two- or three-dimensional structure), versions (salts of the parent, and derivatives such as isotopically labelled substances), and preparations (the result of one particular experiment) is important and the structure of the company registration number and the registry file often reflect the need to preserve this relationship in retrieval and classification.

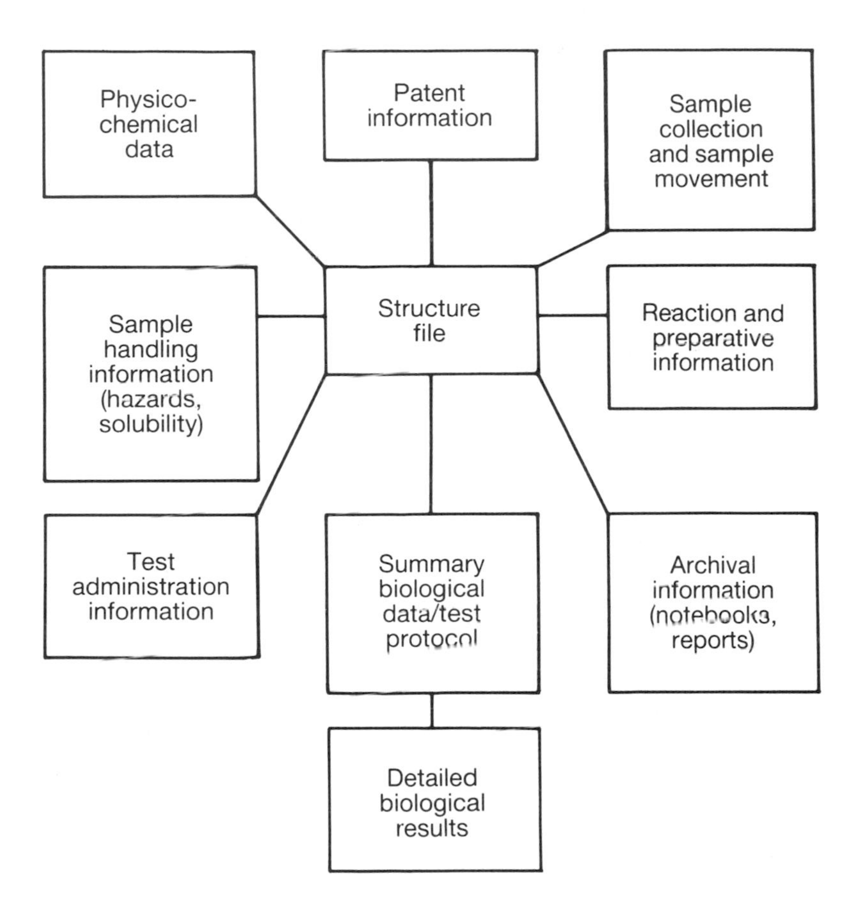

Fig. 1.2 – Components of an internal information system in the pharmaceutical industry.

Structural access to commercially available compounds is also important and the pharmaceutical industry made a considerable contribution to the development of the Fine Chemicals Directory, an online service providing computerized access to chemicals from a range of manufacturers [5, 6].

Although the use of coding systems for structural representation has resulted in registry files which are predominantly maintained and searched by the information scientist, the developments in graphics software described in Chapter 7 are stimulating companies to review present systems and to consider developing and purchasing graphics software [7, 8, 9].

In conjunction with structural information, files of data generated on the compound are also developed. Priority in mechanization has been given to biological test results because of the need to manipulate chemical and biological data for structure activity work. Most biological data handling systems have been designed to meet two needs: the simplification of data recording and communication of test results are almost as important as the creation of databases for retrospective searching [3, 4]. Biological data collection is frequently centred in the laboratory with the scientist entering laboratory results interactively online or with data being captured directly from instruments. Once validated, data become available for searching by users outside the originating department. The immediate accessibility of new information is a signal for discussion and communication between the chemist and his non-chemical colleagues.

As well as evaluated data and a simple summary of activity, files will include information on controls and references to protocols. They may also reference raw data and reviews of information held in notebooks and reports. The size of central biological data files can be enormous – of the order of three million test results within Merck, Sharpe & Dohme. Display of information from such systems will include structural formula and activity data. The chemist will be given facilities for tabulating and reordering information and for reporting information: the use of interactive online systems can lead to a diminution in the formal reporting required within an organization. Increasingly important is the need of the chemist to download information from central systems for use in personal files where it can be manipulated creatively using a range of analytical and statistical packages. Facilities for using internal data in conjunction with data generated externally are required – for instance, use of the Hansch database (see Chapter 4) with internal data to determine structure–activity relationships. Where experimental values are not available, the chemist will want access to programs for predicting data values. Interfacing of internal information systems with data calculation software and with systems for modelling the three-dimensional behaviour of molecules of interest is also growing.

Most pharmaceutical companies have well-developed computerized or manual indexes to company R&D reports (the industrial equivalent of the primary literature). Indexes are also maintained to laboratory notebooks and other archival documentation. The company internal registry number provides a key access point to information from these different sources. Collections of compound samples are maintained and indexes to these include location and amount: these compound samples represent a huge financial investment and it is crucially important that maximum use can be made of them. Reaction information is traditionally less generally mechanized although a very few companies have produced in-house reaction indexes [10, 11] and there is a growing take-up of commercially available software, e.g. REACCS.

Many companies have developed methods for systematic access to physicochemical information. At the Wellcome Research Laboratories, Beckenham, physical chemists can access analogue data on microfilm via a computerized index which allows access by name, fragment code, internal registry number, and molecular formula [12]. Technique and instrument are other retrieval keys and methylene and methine counts are particularly useful in retrieving NMR measurements. Administrative details such as the date of measurement and the scientist responsible are also stored.

This Wellcome database fulfils many of the basic requirements of the bench physical chemist. Its maintenance, though labour-intensive, is no more than the normal premium required for computer-based information systems. However, its greatest limitation is its unsuitability for cross-correlation of data, since data is not stored on the system. Cross-correlation is enormously important since physical chemists cannot normally solve problems by use of only one technique but need the interaction of several techniques in complex situations. Few companies have developed in-house systems for storing actual physico-chemical data values in machine-readable form. With the reduction in cost of data storage, interest in storing this type of data internally is growing.

Handling of published literature in internal information systems is generally limited, by the resources required, to a key literature reference. Within the Basle-based BASIC group of companies, similar techniques for accessing *Chemical Abstracts* and the internal registry files have been available for some time [13] and a major advance is the development of software which at Sandoz allows simultaneous novelty checking of the Chemical Abstracts Registry System and the internal files to retrieve relevant internal and Chemical Abstracts Registry Numbers for a given structure. Some companies will store references to their own patent information within their company files but this is not common [14].

Although the majority of internal data handling development in the pharmaceutical industry has concentrated on primary research and data generated prior to the development of a candidate product, much effort is now being placed in establishing systems to handle the vast amounts of analytical, physico-chemical and other data produced during product development. The techniques developed for manipulation of large volumes of data on a single candidate compound are of no direct relevance to this discussion since they are used more for simplifying administration than aiding creativity. However, they are noteworthy in that they require the development and physical chemists to rely on the computer increasingly for test control and data processing.

The main theme of internal information handling is therefore the systematic and highly controlled documentation of all internal information of long term value expressed and stored in such a way that the real value of the information and the sensible limitations on its interpretation are easily determined. Such information systems although originally used only by information scientists have been developed in close conjunction with the chemist and are now undergoing development to allow the chemist direct, simple and flexible access to data. Data are available therefore in a form which allows their immediate use and does not refer the enquirer to a number of secondary reference points. These achievements clearly distinguish between internal information systems and external systems for published information where the existence of data is frequently difficult to ascertain, where the data itself is difficult to extract without reference to primary publications, and where retrieval software is not marked by its user-friendliness.

1.3.4 The academic research chemist

Many of the information needs of the academic chemist involved in research are similar to those of the industrial chemist. The academic chemist has the same need to be kept aware of current information, to obtain access to retrospective information, and to communicate his own information. The published information tools used by the academic chemist tend to be the more traditional manual ones. Reasons for this are discussed in section 5 of this chapter. Academic chemists regard the scientific journal as their major means of communication and are prime contributors to it [1]. They are, however, often notoriously ignorant of the patent literature in comparison to their industrial counterparts. Unlike the pharmaceutical chemist the academic will not require access to information in the non-chemical or peripheral disciplines to anything like the same extent as the industrial chemist.

Another major difference can be seen in the organization of self-generated information. In this area, the academic researcher is less accountable than his industrial counterpart. The number of immediate colleagues who require access to a particular scientist's data immediately is minimal and there is no corporate requirement to handle data to specified internal procedures. This can be exemplified by looking at the impact of a feasibility study conducted at the University of Sussex to investigate the need for, and problems associated with, the establishment of compound collections in university chemistry departments and the production of manual and machine readable indexes to these compounds. The study showed that present methods of storage and retrieval were haphazard but there was little enthusiasm for change. Few chemists were sufficiently interested to organize their own collections for any benefit of more efficient organization, such as encouraging the sale of samples between universities and industry. Even the cost of synthetic effort was not a motivating factor in the minds of most chemists. In general they would rather duplicate a synthesis and characterize their own compounds even if they knew a sample existed. The general feeling was that 'there is nothing like doing something for oneself' [15]. This specific example may highlight a more general belief that priority should be given to new research rather than recording of previous results. That academic chemists do sometimes see justification in certain instances for cooperation is, however, demonstrated by the development of such cooperative schemes as the Cambridge Crystallographic Database and the work of the Mass Spectrometry Data Group which will be discussed in Chapter 4.

1.3.5 Limitations of existing chemical information systems and services and areas for development

Considerable progress has been made in the development of sophisticated information systems for handling company information and advances such as CAS ONLINE and DARC are providing much-improved methods for accessing published literature. It might therefore be thought that, with such a range of information services at his disposal, the research chemist, particularly the pharmaceutical chemist, could have no further requirements. There are, however, certain deficiencies in current services which can be clearly recognized. These can be discussed under two headings: availability and accessibility.

1.3.5.1 *Availability*

With the enormous volume of chemical publication and the variety of secondary services, non-availability of information might not be thought a problem. Chemists certainly seem happy with the *status quo* in journal publishing and in general, show little interest in developments in electronic publishing and the synopsis journal. In the secondary services, however, there are gaps in coverage of considerable concern.

Within chemistry itself the two major deficiencies are the absence of any good chemical reactions database and the limited availability of data on chemical compounds. Other problems for the chemist are the poor organization of the non-chemical literature for retrieval of chemical information and the lack of simple, structurally indexed services for patents literature.

Reaction indexing

Chemical synthesis is the prime activity of the chemist. The reaction literature is, however, extremely poorly organized for retrieval. Computer-assisted retrieval is available (see Chapter 8) but databases are not comprehensive and are unsuitable for direct use by chemists. Retrieval is limited to the use of text-based dictionaries or fragment codes with resultant false drops or the need for tracking down detailed records, and no simple methods of substructure searching the reaction literature yet exist. Computer-assisted synthesis is an interesting concept but to date the transform libraries are inadequate and the libraries which are just beginning to develop through cooperative groups are unlikely to be made publicly available. The participation of industry in these groups stresses the industrial importance of this type of information.

Because acceptable techniques of reaction indexing have only recently become available, few efforts have been made within companies to develop comprehensive reaction indexes to their own research experience although some do exist [10, 11] and simple records of preparative routes may be collected in central files. The availability of commercially available software for reaction indexing is beginning to make a really significant impact. Ideally, the chemist requires a service which provides structural access to a public collection of reactions while providing facilities for creating a local reaction library.

Data

The chemist's prime interest in seeking information is, in many instances the properties of structures- physico-chemical data, hazards and biological activity. Frequently these data are needed for immediate use; more often than not, unless the data exists within a company databank, they are not directly available. Many data are published in the primary literature. However, indexing is poor and the existence of data therefore difficult to establish in secondary services. Retrieval, both manual and online, therefore requires a laborious tracking from abstracts to papers themselves in order to ascertain whether a data item is included. Often, data are found in handbooks and reference books. Here extraction requires a thorough knowledge of and physical possession of the requisite sources since few reference texts are available in machine-readable form.

Often, once located, data cannot easily be used, since they have not been published to any international standard format and may be found to be expressed inadequately with insufficient context given. Although organizations such as CODATA have made substantial progress, much more work needs to be done to standardize data reduction methods internationally.

Major problems for the chemist therefore are the difficulty of identifying the presence of data in the secondary services and the lack of computer databanks containing good quality data presented in standard format which permits cross-correlation within and between techniques and allows retrieval by graphical structure searching methods. Again, as for reaction indexing, systems which allow both local processing of proprietary data and access to publicly available databanks are required, with facilities for downloading published data to be used locally for analysis and prediction.

Another problem in the area of data availability is that of non-publication. While the criteria of the scientific journal for selection of papers ensure that new science is highlighted [16, 17], this is at the expense of results which may be routinely predicted or result from application of standard procedures or techniques. This has the consequence of reducing the data available in the literature. In addition, within industry, commercial sensitivity will understandably restrict publication of chemical research, and the lack of a vehicle for publication of data on non-sensitive compounds acts as a further barrier. A further cause for alarm is the 'tied' data bank. Much physico-chemical data is available from instrument suppliers, some of which restrict access to purchasers of their equipment. Further discussion is required on systems and procedures whereby scientists can provide quality data on compounds of known structure in a standardized format to a central databank of general accessibility thus encouraging 'publication'. The route should be a direct one, with the minimum of time between data generation and data availability and streamlined validation procedures to ensure accuracy.

The non-chemical literature

Once outside the major secondary services in chemistry the chemist finds that information on chemicals is widely scattered and poorly indexed. Retrieval by structure is difficult and retrieval by substructure impossible. The problems are seen in the 'fringe' interests of both organic, inorganic and physical chemists. Improvements could result if the abstract services for chemistry widened criteria for selection or if abstracting services for other sciences standardized and improved their methods for describing and indexing chemical structures. The ideal would be the unification of access techniques to secondary services.

Patents

Crucial to industrial work is the determination of novelty for a product candidate so that patent protection can be ensured. The adoption of the generic (Markush) formula to exemplify the structures covered by a particular patent presents particular hazards to those attempting to determine novelty. Although patents are abstracted in *Chemical Abstracts*, the Derwent CPI Services, INPADOC and ISI services, no entirely satisfactory services exist. Because of the financial importance of this area it is imperative that commercially applicable methods for identifying the substances covered within a patents claim are developed and adopted for patents information services, also that these methods provide a novelty and substructure searching route which is simple enough for the chemist to use directly.

1.3.5.2 *Accessibility*

Printed tools are still the main tool of the chemist who undertakes his own information retrieval and, at the moment, the prime user of computer-based chemical information systems is the information scientist. Obviously, while effective exploitation of such systems requires knowledge of a large number of sources, many computer operating systems, several query languages, and a series of routines for access, as well as keyboarding skills and knowledge of telecommunications, the information scientist will continue as an essential intermediary.

Two factors, nevertheless, support the prediction that increasingly the laboratory chemist will wish to acquire information directly from external and internal databases [18]. One of these is the increasing involvement of chemists in the use of computers for experiment design and control and laboratory data management which will overcome the problem of familiarity with terminal operation. The other and more significant factor is the increasing availability of software which, for the first time, allows the chemist to carry out a dialogue in his own language, that is, the language of the structural formula. Graphics-searching software allows the chemist to draw a structure or substructure on the screen of a graphics terminal and provides an answer as an attractive structure display, which in some systems may be supplemented by associated information in the form of data and text. The opportunity to interact with databases in this way can, if properly exploited, provide the bench chemist with a more effective tool than the most efficient intermediary, promoting generation and development of ideas. Also, time will be saved if the chemist can undertake retrieval, and this is possible if easy and reliable routes are available. It is noticeable that the marketing of services such as CAS ONLINE and DARC and software such as MACCS and REACCS is heavily towards the chemists themselves.

The bench chemist must therefore be regarded as the target customer for future information services. What are the existing problems which may hinder access to information and which must be resolved in the future?

Integration

The first and a major problem which has been alluded to throughout this chapter is the fragmentation of existing services. Although the chemist has many information services available to him, different access routes are required for each, and therefore comprehensive enquiry may be extremely tedious.

The chemist wishes to search for full structures and substructures graphically with the option to include stereochemical features. In answer, he requires data and references on this structure, displayed together. Textual searching is important; the chemist will wish to search keywords, authors, trade names, titles of books and journals and combinations of these. The combination of textual and structural search terms should be possible. The importance of data has been stressed throughout this chapter and data must also be retrievable and searchable using online techniques. Access to patents and reactions is also required. Not only does the chemist require to access and search this information but he requires a single access point through which to gain entry to a universal chemical information system. In addition, the industrial user of information systems wants both to search the published information and access company data in a similar way [19].

The goal of the fully integrated system may seem difficult to attain but it is certainly feasible. One possible route might be to use the CAS Registry Number as a connection between all the elements of the system. This, of course, depends on CAS Registry Numbers being available to everybody at a reasonable cost. Alternatively the development of graphics/data/text retrieval software which might serve as a common front-end to a series of differently structured but essentially connection-table-based systems should be achievable. Active discussion by both chemists and information scientists is required to achieve the integration required.

Retrospective coverage

Access via computerized techniques is only really available for literature published since the mid-1960s. Chemical and physico-chemical data can in contrast be valuable even if it was produced in the nineteenth century. Further developments in chemical information handling must therefore ensure that key older collections are computerized and the Chemical Abstracts Service pre-1965 registration project is particularly welcome in this respect.

The simple interface

The term 'user-friendly' is becoming somewhat hackneyed. The principles it embodies are nevertheless extremely valuable. Graphics search systems are available – they are not yet simple and foolproof. Improvements in substructure searching capability are required particularly in the specification of stereochemistry. Also, simpler methods of expressing generic structures with a range of possible substituents and an easy route to formulating searches for a range of homologues is needed. Particularly for the end user, search construction should be as simple as possible and not only should improvements in substructure searching be made, but the existing methods of text searching should also be reviewed.

Much effort needs to be devoted to the development of online search aids for the chemist which allow him to express a query in his own terms which can then be converted into the required search expression for the particular system to be accessed.

Access to reference information

In discussing the integration of information systems, the storing of all key data required for searching and display was envisaged including abstracts of the published literature. While this will satisfy many needs, if the scientific paper continues as a prime method of communication, logically it too will have to become part of this ideal system, capable of being retrieved and browsed online, not only for retrospective work but also for current literature activity. This development will undoubtedly happen, although whether the online journal will ever become a significant tool for the end user cannot yet be predicted with certainty.

Currency

In general, the delays in production of publications have bred a tolerance which is at variance with the need of the industrial chemist for immediate access to his own data and the data of colleagues. Certainly the use of information technology offers the opportunity to streamline and integrate primary publication with database and databank creation. Time taken to publish will be greatly reduced, as will costs, when data can be directly transmitted to central databanks and when the initial typing of a document provides a direct input into the start of a computerized publication process in which the role of referees and editor can be retained but the production process greatly streamlined.

Costs

A detailed analysis of the relationship of cost to benefit of chemical information services falls outside the scope of this book. Two points should be made: (1) Accessibility of information systems to chemists is related to cost and there is clear evidence that organizations, particularly the universities, are being forced to review most critically budgets for published information. (2) Industry too is attempting to ensure that only information of real value is purchased and management requires sound justification of internal information systems development in terms of reduced administrative costs and increased activity.

Given these factors, the costs required to develop a fully integrated chemical information system might seem to be difficult to justify. However, investigations in industry [19] have shown that up to 20 per cent of experimental work could be saved if all the available information were fully used. This represents a real benefit of significant value. The same groups have predicted that the use of a complete and decentralized online system would increase by 10–100-fold of the present level. This again represents an increase of real magnitude sufficiently to justify a real increase in investment in future information services.

Communications

The chemist will have much less tolerance than the information scientist when faced with computing and telecommunication failures. Reliability of communication channels which are regularly open and cannot be disturbed or interrupted by technical failures is therefore regarded as an essential improvement to existing services.

1.4 INFORMATION FOR CHEMISTRY TEACHING

The purpose of an academic institution is to create highly educated and/or trained people. Many of these will pursue their vocations in industrial or other research laboratories, joining the large existing user body of research chemists towards whose needs the efforts of chemical information providers are mainly directed.

The information requirements of the academics who teach are somewhat wider than those of chemists involved entirely in research. The chemistry lecturer fulfils two main functions: firstly, he meets the different demands of teaching non-chemists, chemists, and research students, the balance between these various activities depending on the type of institution to which he belongs; secondly, he critically prepares and then delivers his teaching material to students, devises and implements assessment procedures to test whether the students have assimilated the contents, and carries out a wide-ranging and time-consuming range of associated administrative tasks. In addition he will be expected to maintain some research activity.

In his training capacity, a teacher in tertiary education may need to cover five classes of material:

(1) Information in his own area of speciality dealt with at an advanced level.
(2) Familiar, high level material with which the lecturer is well acquainted.
(3) Straightforward material for service teaching to non-chemistry specialists.
(4) Novel material for special topics with which the lecturer is conversant only in part.
(5) Fairly simple but unfamiliar material required as courses are re-allocated.

The information requirement of the first category is covered by the academic's own research activity and should therefore represent no real problem since the teacher will have access to as wide a range of published literature as does the industrial researcher. In the other areas, a series of problems present themselves, all of which might be solved by the development of three new services.

1.4.1 Critical bibliographies of textbooks

The chemist is supported in teaching the material described in classes (2) and (3) by many standard texts. Here, however, lies the danger that, having established particular sources, familiarity and natural inertia may hinder critical review of these and prevent the selection of newer and better alternatives. Although book reviews are available they are scattered through many publications and are normally published prior to real use of the text. What is therefore needed is a comprehensive and critical annual bibliography. This would be compiled from book reviews and publishers' announcements – the printing of these in juxtaposition would be both informative and stimulating. The bibliography would be strengthened by the indication of assessments collected from academics with practical experience of the text. Such a tool would be invaluable not only for the academic teacher but also for the industrial chemist who will also require reference books of quality when entering new areas.

1.4.2 Improved access to review literature and the development of non-specialist reviews

The majority of chemistry degree courses incorporate some advanced work on a selection of special topics in the final year. Ideally, such topics should be taught by specialists. However, from time to time a non-specialist is required to teach at a high level a subject with which he is unfamiliar. This can be particularly true at a time of industrial recession when new recruitment into the academic sector is reduced and new research areas are not actively taken up by older staff whose interests have become static.

The lecturer needing to present a course in a new field requires a book or article that offers an unbiased survey of the whole area, starting from a level appropriate to a non-specialist. Because of its comprehensive nature, a book is usually the most helpful source. A rapid scan through the contents normally enables the practised lecturer to select the key portions on which a balanced set of lectures can be based.

Reviews are considerably less useful. This may seem paradoxical. However, the tertiary literature generally has become more suited to the researcher than the educator. The main reason for this is the enormous growth in chemistry. In consequence, an author cannot cover a topic from its origins and also fit into a review of restricted length a detailed appraisal of the most recent developments. Authors are experts in their reviewed field and they tend to emphasize the aspects which are uppermost in their own minds. Usually these are the most exciting advances, which often happen to be the most recent. The result of this is that reviews are more suited for updating than overviewing.

The educator requires a summary of a subject that is by, but not for, a specialist. This type of review would also be useful in industry where from time to time specialists are asked to change areas and could therefore utilize this material most effectively. Two specific actions are required: the establishment of a new abstracting category, 'the E category' and the introduction of a new series: 'Non-specialist periodical reports'.

1.4.2.1 *The E (educationally useful) category*

The educator needs to identify those reviews which are overviews rather than updates for the specialist. Indeed, the coverage that the academic chemist needs may lie not just in the primary literature, but in theses, or even in manufacturers' technical handbooks, and these are currently difficult to identify.

What is required is a new abstracting category, within, for instance, *Chemical Abstracts*, to augment the existing R and P categories. Alternatively a new secondary publication, either in hard copy or online should be considered. Material to be assessed in this way would cover all the major aspects of the topic, not just the most recent, and would contain sufficiently well described references to enable the user to follow up particular aspects in more detail. Material would be captured from the widest possible range of sources. Obvious candidates for the classification would be many of the articles in chemical education journals.

1.4.2.2 *Non-specialist periodical reports*

Although a service identifying published non-specialist reviews would come some way to meeting the needs of the educator, it would not remove the problem that insufficient reviews of this type are published. A new overview publication is therefore required with the following objectives:

(a) to cover areas of present or growing interest;
(b) to provide a critical and balanced coverage, starting from a position, and a level, appropriate to the non-specialist;
(c) to present an unbiased view of rival theories and interpretations pertinent to the field;
(d) to survey the past, present and likely future developments in a rational and brief manner;
(e) to offer informed guidance to the reader who wishes to pursue particular aspects of the field in greater detail. For a service of this type a critical and annotated reference list would be essential.

It is likely that this new tertiary vehicle could be launched under the auspices of a major learned society. Since the need for dependable and comprehensive overviews is worldwide, it might be appropriate to explore the possibility of collaboration with learned societies in several countries. Bearing in mind that the Third World also has a need for texts assisting in rapid and accurate self-education, agencies such as UNESCO and IUPAC might have a role to play.

The price of the new publication would need to be low to be affordable throughout the user community.

1.5 TRAINING STUDENTS IN INFORMATION USE

Straw polls of most collections of industrial chemists reveal, both in the UK and to an extent in the USA, little exposure to methods for using and searching published literature in their undergraduate and postgraduate training, still less in methods for systematically organizing their own information. This somewhat subjective measure is borne out by the wider experience of the British Library Chemical Information Review Committee [1] who concluded that

(a) many chemistry departments have no chemical information course;
(b) courses, if available, are not subject to examination and consequently tend to be regarded casually by students;
(c) courses tend to be regarded as uninteresting;
(d) courses are very locally based, that is directed to use of local library facilities rather than a more general guide to information services;
(e) courses are frequently taught by a lecturer who is neither especially interested nor well-informed about chemical information retrieval.

There are notable exceptions: paper 4 of the GRSC course in the UK requires substantial coverage of information techniques, but,. this course is taught predominantly to the non-university sector. In the USA more effort has been made to establish this type of training but undergraduate training is far from ideal. Most postgraduate course work includes some tuition in information techniques but this is rarely sufficient to ensure real expertise. Some universities have appointed information scientists to train chemists in information retrieval as well as servicing chemical departments [20]. Such appointments, are, however, relatively rare.

The lack of formal training in information work is exacerbated and certainly partially caused by the lack of penetration of modern methods of information handling, for instance, online retrieval techniques, into tertiary education establishments. In the UK this is partly due to financial restrictions since little provision is made for purchase of information within research grants. However, this is not the only reason. Academic chemists are generally insufficiently aware that the handling of information is now an essential skill for the practice and teaching of their subject. The lack of importance attached to knowledge of information handling and retrieval may partly stem from the exclusive position of leading academics whose 'invisible college' will keep them informed of the latest developments and who are therefore no longer exposed to the need to keep up to date in a more personally active way. The academic chemist too is under less pressure than his industrial counterpart to make rapid and effective use of the most recent literature. This propagates a situation in which the student with whom the academic is in contact is guided down the wrong path, if he is in fact guided down any path at all. This must change. Academic chemists must increase their own awareness of primary, secondary and tertiary information services. Otherwise, their ability to participate in an informed way in the teaching of information skills will vanish and they eventually will not even be able to converse intelligently with students, who have been trained in these techniques by other hands. Their own research skills will also suffer.

More effort is made within those industries which have well-developed library and information functions and where librarians and information scientists are available to introduce and encourage chemists to make active use of their local information services. It is considered essential that the industrial chemist is encouraged to undertake searching which he can do quickly, simply and precisely and equally essential that he be encouraged to delegate more complex searches which require much more time and/or the use of online services to the information specialist. Since techniques for information retrieval change and new services develop, industry will upgrade and extend the chemist's knowledge as appropriate. Also within successful companies, well organized procedures for internal information handling will exist, progress will have been made in the computerization of internal data, and the chemist will be not only compelled to participate in general documentation systems but will be encouraged to make individual data and documentation systems as efficient as possible.

The training which industry undertakes, however, should only need to build on good undergraduate and postgraduate training in centres for tertiary education. It is therefore extremely important that the attitude to information storage and retrieval tuition within university chemical departments begins to approach the enthusiasm of industry. Some priority must be given to the inclusion of information skills in the undergraduate curriculum, preferably taught by chemists to reinforce the importance of these issues. A training scheme should include the following:

(a) production and use of primary journals;
(b) the importance and nature of major chemical abstracting tools such as *Chemical Abstracts* and the abstracts literature of related sciences;
(c) patents and patents information services;
(d) reference books and data banks;
(e) theses, dissertations and reports and 'grey' literature, and abstracts services covering these;
(f) online databases and vendors;
(g) current awareness techniques and services;
(h) retrospective search procedures including online searching;
(i) personal information handling: laboratory documentation, report writing, indexing, the use of computers for storage, retrieval and manipulation of data;
(j) the reference literature on information sources and retrieval techniques.

At postgraduate level, experience with use of the services must continue and the postgraduate researcher must be encouraged to keep up to date, not only with the scientific information relevant to the project but also with developments in information sources and techniques within his scientific discipline.

Ideally, chemical information topics would be presented to the student by a team of academic chemists and information scientists who also should preferably have a chemical background. This would overcome the problem that the former represent informed users, but their knowledge may not be comprehensive, whereas the latter know about the facilities that are available.

In addition to the above recommendation, three more can be seen as being desirable:

- organizations such as learned societies must offer accessible courses to assist in this training;
- the bodies responsible for funding research must be more supportive of costs for using modern information retrieval tools when awarding research grants and postgraduate support allocations;
- consideration should be made to limiting full membership of professional chemical societies to those who can display competence in chemical information techniques.

The problems to be overcome in achieving progress are not insignificant and require considerable effort both from educationalists and information scientists.

1.6 THE ROLE OF THE INFORMATION SCIENTIST

Both in providing and using information it would be far too optimistic to assume that all chemists thoroughly appreciate the need for documentation and always take time and effort to ensure that they acquire the right information at the right time. The information scientist provides a stimulus for good information use and is generally accepted as helpful by those chemists in industry and academic life who can use a formal information service.

The two traditional roles of the information scientist are:

- to encourage the chemist to use available information and to recognize the stages in research where this is crucial;
- to ensure that the chemist uses the simplest and most effective route to acquire information; this can include a catalytic role in promoting interaction between people in a company who are ignorant of one another as well as acting as an intermediary in retrieval of published information.

It is, however, important that the information scientist continues to fulfil these responsibilities only if he can undertake to do so in such a way that no barrier is placed between the chemist and information sources. The chemist is now ready, anxious and capable of using online services and involvement of the chemist in computerized information retrieval is a must. In some companies, training courses have already been undertaken although with mixed results [21, 22].

If the chemist takes back responsibility for the majority of information retrieval, what then happens to the information scientist?

In the first place, the information scientist will still undertake enquiries for the infrequent users, for instance senior management. Also there is evidence that currently where chemists become trained in online searching techniques they quickly recognize the need for information scientists to undertake difficult and complex enquiries, thus increasing rather than lessening the pressure on the information service. Certainly until integrated services are available and where, for example a substructure search has to be combined with patent information and text searches in order to answer a particular enquiry it is unlikely that the chemist could be more efficient than the information specialist.

The chemist will continue to need guidance in the selection of the most effective tools both for manual and machine retrieval. The role of an information scientist as a specialist in knowledge of information sources and the structure and content of databases will therefore remain crucially important indefinitely. The information specialist will also be required to support chemists in their use of information services by training, on-the-spot advice, and by general encouragement of the development of information handling skills of end users. Simple guides will be required and the information scientist will, as well as being influential in selecting hosts and databases of most use to the end user, be responsible for keeping up to date with the flood of material emanating from the database vendor. These services will be equally essential to the industrial and the academic chemist [18, 20].

One other key role for the information scientist will be a continuation of his involvement in systems development. The information scientist has been a prime stimulus in the development of internal industrial information systems and external services. He must therefore recognize a continuing responsibility to encourage the making available and systematic use of information through the design of good internal information systems. In addition the information scientist must exert pressure for improved access to externally generated information. This will require not only lobbying publishers but also the encouragement of chemists to cooperate in the exchange of information, in the agreement of standard data formats, and in full and systematic documentation of research results. The information scientist must also lobby actively for and with the chemist on information matters, for better education in information storage and retrieval, for generation of new services, for the development of the ideal chemical information system. Last but not least the information scientist must be the first to see the opportunities created by the latest developments in information technology, must highlight these to the chemist, and must encourage the chemist's active commitment to improve existing services. To do this effectively the information scientist must continue to make every effort to fully appreciate the working environment and problems of the chemist.

1.7 CONCLUSION

Formal attempts to review the information needs of chemists have been few and published references are negligible. Although not fully representative of the entire chemical community, this introductory chapter has attempted to summarize the views of chemists on current information provision and future information needs. Overall, the needs of researchers are remarkably similar between industry and the university although the specialized requirements of the academic chemist in the teaching role are exceptional.

The major differences between the academic and industrial research chemist centre on the costs of modern information techniques and the availability of money to purchase them. Certainly in the UK, financial problems are creating a two-tier chemical information society. In the context of this book such a situation can only be raised as a problem. However, it is clear that within higher education and academic research the use of online published information systems is growing very slowly in comparison to industry. It is a paradox worthy of further discussion elsewhere, that centres of fundamental research are likely to continue with very poor information provision, hindering their efforts to fulfil their role.

The need for more and greatly improved training in information techniques must be accepted by the chemical profession and the recommendation that information handling training be incorporated into all chemistry courses is extremely important.

For the research chemists, the paramount need is the integration of all available sources of chemical information or the development of common access techniques to these. The need for this utopia in services will increase as chemists become more involved with searching the commercial databanks of published information, since the end-user has less patience and less time then the information intermediary.

If this universal interface to all relevant information can be achieved the chemist will have available, from his terminal, all possibly relevant information at his finger-tips. Only then can the enormous potential of the developments which have taken place so far and which will be described in the remainder of the book be realized.

Since all information services only exist to serve the needs of a customer population it is essential that the information industry refocuses on the chemist and allows the chemist to exert real influence on the information systems essentially designed for his benefit. The chemist has already recognized his alienation and lack of influence on the development of services and is unhappy at it. The chemist has the most important contribution to make both in services developed externally to his organization for published literature and those developed within his organization for handling his own information.

REFERENCES

[1] Rowland, J. F. B., Information Transfer and Use in Chemistry, Final Report of the Chemical Information Review Committee, *BL R&D Report No. 5385*, March (1978).

[2] Almond, J. R., and Welsh, H. M., Chemical Substructure Searching – Industrial Applications and Commercial Systems, Drexel Library, Quarterly, **18**, 84–105 (1982).

[3] Howe, W. J., Milne, M. M., and Pennell, A. F., Retrieval of Medicinal Chemical Information, *ACS Symposium Series No. 84, Washington, ACS* (1978).

[4] Becker, J., Jung, D., Kalbfleisch, W., and Ohnacker, G., CBF – Computer Handling of Chemical and Biological Facts, 2, *J. Chem. Inf. Comput. Sci.*, **21**, 111–117 (1981).

[5] Walker, S. B., Development of CAOCI and its use in ICI Plant Protection Division, *J. Chem. Inf. Comput. Sci.*, **23**, 3–5 (1983).

[6] Rosenberg, M. D., Debardeleben, M. Z., Debardeleben, J. F., Chemical Supply Catalog Indexing: Now and the Future. An Ideal Place for the Use of Wiswesser Line Notation, *J. Chem. Inf. Comput. Sci.*, **22**, 93–98 (1982).

[7] Howe, W. J., and Hagadone, T. R., Molecular Structure Searching: Computer Graphics and Query Entry Methodology, *J. Chem. Inf. Comput. Sci.*, **22**, 8–15 (1982).
[8] Hagadone, T. R., and Howe, W. J., Molecule and Sub-structure searching: Minicomputer Based Query Execution, *J. Chem. Inf. Comput. Sci.*, **22**, 182–186 (1982).
[9] Polton, D. J., Installation and Operational Experiences with MACCS (Molecular Access System), *Online Review*, **6**, 235–242 (1982).
[10] Zeigler, H. J., Roche Integrated Reaction System (RIRS). A new Documentation System for Organic Reactions, *J. Chem. Inf. Comput. Sci.*, **19**, 141–149 (1979).
[11] Osinga, M., and Verrijn Stuart, A. A., Documentation of Chemical Reactions IV. Further Applications of WLN Analysis Programs: A System for Automatic Generation and Retrieval of Information on Chemical Compounds (AGRICC), *J. Chem. Inf. Comput. Sci.*, **18**, 26–32 (1978).
[12] Ward, S. E., Integrated databases for chemical systems in *Proceedings of the CNA(UK) Seminar, Integrated databases for chemical systems, Canterbury*, pp. 47–73 (1979).
[13] Graf, W., Kaindl, H. K., Kniess, H., and Warszawski, R., The Third BASIC Fragment Search Dictionary, *J. Chem. Inf. Comput. Sci.*, **22**, 177–181 (1982).
[14] Johns, T. M., and Ryno, D. I., Patent Searching in a Pharmaceutical Company, *J. Chem. Inf. Comput. Sci.*, **18**, 79–80 (1978).
[15] Stevenson, H. B., and Conway, A. J., Indexing of Structural Data on Compounds in British Universities, *BL R&D Report No. 5355*, June (1977).
[16] Rhyner, P. (1982), *J. Chem. Soc.*, Notices to Authors, No. 1 (1968).
[17] *Phytochemistry*, Instructions to Authors, **18**, 4 (1979).
[18] Haygarth-Jackson, A. R., Online Information Handling – The User Perspective, *Online Review*, **7**, 25–32 (1983).
[19] Rhyner, P., The international importance of CAS, 184th National Meeting, ACS, Kansas City (1982).
[20] Ockenfeld, M., The Role of an Intermediary in a University Department, 5th International Online Information Meeting, London, Oxford and New Jersey, *Learned Information*, 307–313 (1981).
[21] Haines, J. S., Experience in Training End-User Searchers, *Online*, **6**, 14–19 (1982).
[22] Walton, K. R., and Dedert, P. L., Experience at Exxon in Training End-Users to Search Technical Data Bases Online, *Online*, **7**, 42–50 (1983).

2

Classical methods of communicating non-structural chemical information

2.1 INTRODUCTION

Communication of chemical information dates back to the beginnings of chemistry. The alchemist of the Middle Ages passed on his information verbally: the apprentice learnt from his master and his learning was, in turn, passed on to his successor. After word of mouth transmission came pictorial, iconic representation giving rise to the first textbooks around the fifteenth century. Textbooks provided one of the primary means of carrying chemical information from the beginning and have continued to the present time but with slightly different functions. In the early days of textbooks, there were no theories and chemicals were referred to by name, with shorthhand symbols to describe the names. The turning point came with the gas laws of Avogadro and Boyle and then, in 1860, modern structural theory was evolved following the discoveries of Dalton, Berzelius and Cannizzaro. The words *chemical structure* were used for the first time and the distinction between structural and non-structural chemical information could be made.

The traditional area of non-structural chemical information communication is well documented and the methods used are similar to those employed for searching for textual information in other scientific disciplines. There are several excellent textbooks both on general scientific information resources and on the specialized chemical literature for the reader who wishes to make a detailed study of the subject [1–6]. A brief review of the traditional methods is given in this chapter because although most of the book is concerned with the technological advances and changes taking place in the area of chemical documentation, the recent developments in online searching stem from the mechanization of the traditional methods of information gathering and it is important that these basic methods are understood. In addition, the scientific journal is the key to all present-day chemical information communication and is likely to remain so for a long time to come. Although the journal has been in existence for over three hundred years, it has changed relatively little. The growth in volume of literature and economic constraints have led to some changes in the format of the journal and these changes are discussed in the first part of the chapter. The chapter is divided into primary, secondary and tertiary publication. The journal, produced mainly by the academic scientist, and the patent, which provides the initial publication of much of the industrial chemical research, are the most important forms of primary publication and it is on these that the secondary and tertiary literature is based. It is therefore appropriate to devote a large proportion of the chapter to a discussion of journals and patents. The importance of journals is recognized by chemists, but scientists frequently make little use of patents as an information resource in spite of the wealth of information they contain. Some alternatives to the conventional journal have been proposed, but none of these alternatives has significantly affected the publication of the scientific journal. Other forms of primary publication, including reports, theses and conference papers are briefly discussed.

Secondary publication is obtained by the organization of the primary literature into abstracts and indexes. The publication of abstracts and indexes by the secondary services is described, but details of the techniques used in the production are not covered, and references [7] and [8] should be consulted for further information. The mechanization of the production of abstracts and indexes, which has given rise to the online industry, has made a substantial impact on the chemist's use of the traditional methods of searching for information through the secondary sources.

There are several definitions of the tertiary literature, but for the purpose of this chapter it is defined as *evaluated* literature based on both the primary and secondary sources, and includes reviews, handbooks and encyclopedias. The chemist's use of the tertiary literature has been relatively unaffected by the online services. The production of many of the tertiary sources has been mechanized, but the printed volume remains a valuable reference work for the chemist and some of the major handbooks are described at the end of the chapter.

2.2 PRIMARY PUBLICATION

The primary literature includes original published papers (journal articles), patents, theses, reports and conference papers in addition to unpublished material such as laboratory notebooks and correspondence.

2.2.1 Original published papers

The scientific journal began as a multi-disciplinary publication which was usually produced by a learned society. *Philosophical Transactions*, published by the Royal Society in the middle of the seventeenth century, was one of the earliest scientific journals. The first journals merely published the texts of papers which had been presented at meetings of a society in order to reach those members who had been unable to attend. Many journals perform this same function today. Other journals, especially in Germany, were founded by an individual scientist who was the first editor; for example, Hoppe–Seyler's *Zeitschrift* and Leibig's *Annalen.* Gradually, however, during the nineteenth century, the system of primary journals was modified and journals became subject-orientated. Authors submit papers and the editor sends them to referees, in the light of whose comments the editor decides whether or not to publish.

2.2.1.1 *Growth in the literature*

Since the nineteenth century, the enormous increase in scientific activity has fuelled an information explosion with which all users of the scientific literature are familiar. The growth rates of large areas of knowledge have been studied and recorded by Price [9]. Different patterns are observable, however, in the growth in literature of small specialized subject fields. Sometimes a discovery is made and the techniques or theoretical background then available do not permit exploitation. The growth pattern of the literature relating to such a discovery shows an early peak followed by a period of relatively low activity before the early levels are reached and surpassed. This growth pattern was observed for the literature of liquid crystals [10], where the early peak of publishing activity was not surpassed for nearly 50 years. A new field which is immediately recognized as important either from a theoretical or commercial viewpoint will show about five years of extremely rapid growth, characterized by a doubling time in the number of papers produced of less than a year, before stabilizing into a decade or more of exponential growth, with a relatively low doubling time. The literature of interferon falls into this category as shown in Fig. 2.1, where an initial 10-month doubling time developed from 1971 to 1981 into a constant 6.5 years [11].

2.2.1.2 *Change in language of publications*

Germany was the leading scientific nation up to the beginning of the twentieth century, but after World War I, Germany's role declined, giving way to the increase in scientific literature produced in the USA. From the data presented by Narin and Carpenter [12], it can be seen that the fraction of abstracts appearing in *Chemical Abstracts* (*CA*) emanating from Germany and the USA had equalized by about 1938 [13]. During the past twenty years there has been a polarization into the two most common languages, English and Russian, with at present 63 per cent of the journal literature abstracted by *CA* in English, 21 per cent in Russian and only 16 per cent accounting for all the remaining languages [14]. Just over one-third of the English language papers are prepared in countries where English is not the official language, providing striking confirmation that English has become the *lingua franca* of science.

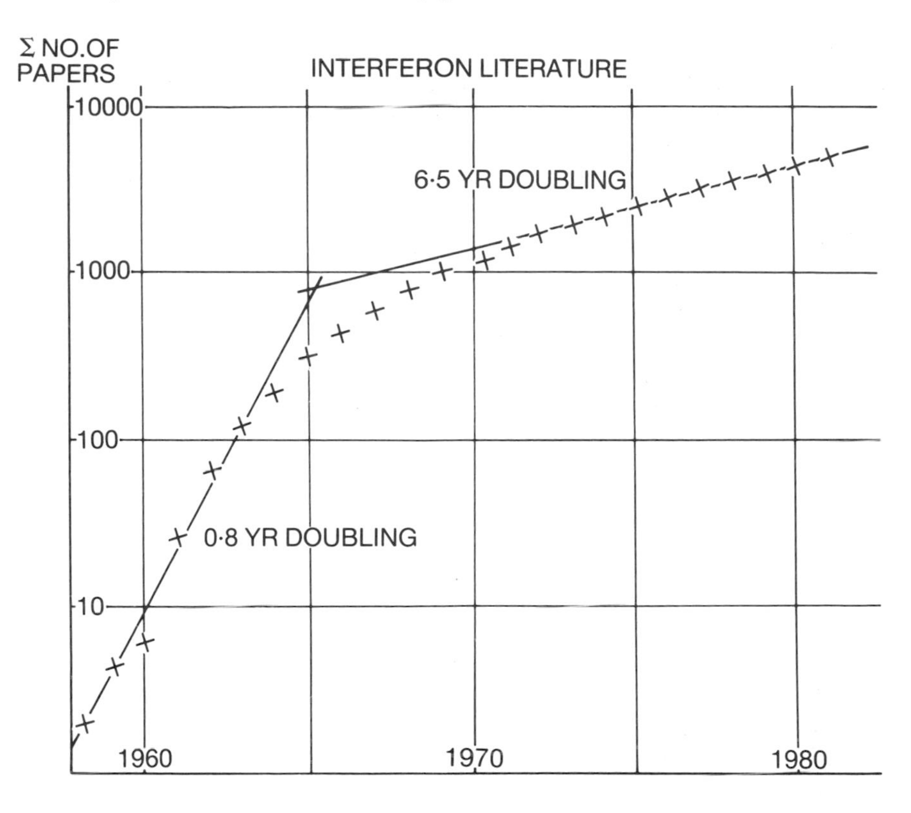

Fig. 2.1 – The growth in volume of interferon literature.

2.2.1.3 *Changes in form of primary publications*

The printed journal has existed in its present form for a very long time. However, minor changes have occurred, mainly for economic reasons. Papers published before 1914 are noticeably easier to read than currently-produced papers. The layout of the older papers is more spacious, page sizes are smaller and the print larger. There is now considerably more information per printed page and a greatly increased number of pages per volume, which has given rise to a very high inflation of journal prices in recent years [15].

Sardar [16] investigated the change in readability between papers published before World War I and the present day. There has been a substantial decline in readability from 1850 to the present day, with the use of more complex language and changes in format and layout. The exponential growth in the number of papers and the economic constraints have caused this change. Some variations between individual journals have been observed [17], but from 1900 to 1920, there was a general increase in the annual number of papers per journal and also in the number of words per paper. Overall, there was an exponential growth in the number of characters per journal. There are limitations to the extent to which type size can be reduced and it appears that printing density in terms of characters per square centimetre has already peaked and there is little prospect of further reductions being made.

A further change that has taken place since the early part of this century is the tendency towards the use of longer and more informative titles [18], probably brought about by the advent of KeyWord In Context (KWIC) indexes of titles in the mid-1950s.

The major changes in communication via the printed journal which have taken place this century are thus:

(a) A considerable decrease in readability.
(b) A rise in the use of English as the preferred language of scientific communication.
(c) An exponential growth in average journal size, which is now showing signs of levelling off.
(d) A small increase in the average length of papers.
(e) The use of increasingly informative titles.

2.2.1.4 *Functions of the scientific journal*

In spite of the minor changes in form of the journal which have taken place this century, the basic function of the journal has remained the same: to serve as a foundation for the advancement of science. Ziman [19] and Ravetz [20] have studied the functions of the journal that have resulted in its conservation during the evolution of the scientific community and have identified four main functions: current awareness, archival storage, quality control and author recognition.

Although journal scanning is valuable for current awareness, its use has tended to be over-emphasized by information scientists. Even in the pre-online era, the journal paper was little used for current awareness [21]. Conferences, seminars, informal visits, letters and telephone calls between laboratories reported research work in progress. The formal paper was usually written on completion of a project, and publication delays meant that the news was stale before the journal appeared. Further delay ensued before the reference was included in the abstracts journal, and thus the journal article could not be an effective current awareness tool in any rapidly evolving field. Garvey [21] showed that scientists obtained their current information by more informal means which have become known collectively as the 'invisible college'. Despite this, formal papers are still written and read, which indicates their worth for purposes other than for current awareness.

In a survey carried out in 1963–64 on behalf of the Advisory Council on Scientific Policy [22], original papers were rated highly for obtaining specific information, particularly amongst chemists with Ph.D. qualifications. A further survey in 1980 [23] showed that academic scientists in particular use traditional scanning of journals to keep aware of the current *published* information in their field of interest. The journal is highly valued because it is often the only source of numerical data, which is invariably not included in an abstract or other secondary publication.

2.2.1.5 *The publishing system*

Leading scientists frequently complain of excessive publication and there are abuses of the publication system whereby scientists publish the same work in different journals or fragment the work excessively to obtain a large number of publications. However, such abuses are rare and the journal publishing system does provide a very subtle means of quality control, which is essential if good, reliable scientific data and valid theories are to be generated. Journals do not simply accept good work and reject bad work. All papers submitted to journals are refereed by a team of scientists with long-standing experience in the field. The same scientist may be asked to referee papers by different journals. He has in mind an informal but well understood 'pecking order' of journals and thus may accept a paper when refereeing for a lesser journal while rejecting the same paper for a better journal. Authors, too, are well aware of the quality of the journals to which they submit their work. The range of journals provides a subtly graded quality control, with the best work appearing in the better journals, down to the poor work in the less respected journals, but everyone's work is available thus providing a complete archive.

Referees and editors may actually improve papers submitted by authors. They are not able to contribute extra laboratory results but may be able to suggest one or two more experiments that would provide extra data to strengthen an argument, or to suggest ways to improve clarity, such as presenting the data in graphical or tabular form. Editors perform a particularly valuable service for authors who are not writing in their native language.

Most academic scientists think of themselves foremost in the role of authors, not readers, of the scientific literature. Scientists need and wish to maintain the system of publishing whereby they receive a reward, both financial and intangible, for recognized good work.

For an editor, the major concern must be to maintain the standards of the journal. An editor must satisfy his authors first and foremost. The price of inappropriate innovation is so high (the downfall of the journal) that it is not surprising editors have proved very conservative. The pace of innovation in primary scholarly publishing will inevitably be slow.

There is, surprisingly, no great economic incentive for change. Even in times of recession, the major British scholarly journals remain profitable and many new journals have been established, often in very specialized fields and with low circulation. The new journals are competing with the established journals for static or declining library budgets, inevitably leading to falls in circulation, and further harm may be done to circulation by relatively liberal laws on photocopying. However, there is no financial incentive for publishers to cease producing the printed journals, despite the fact that the 'electronic journal' is becoming cheaper than the printed version. The Royal Society survey showed that the majority of newly-formed British journals become financially viable within three or four years and very few have ceased publication for financial reasons [24].

The scientific paper is the ultimate source of a substantial proportion of the information that is stored in the sophisticated secondary services. The data from chemical research laboratories form the foundation on which the chemical information systems stand. The majority of workers in the laboratories wish to continue to publish in the way they are used to and thus the printed journal is likely to remain in its present form for some time to come.

2.2.1.6 *The key chemical journals*

Garfield has made a detailed study [25] of the frequency of citation of journals, from which it is possible to identify the major chemical journals, although frequency of citation is not necessarily the key to a journal's importance. In Garfield's survey, only journals covered by the *Science Citation Index* (*SCI*) were examined. He found that one half of all references cited 152 journals, and these 'key' journals were listed in rank order of frequency of citation. Some important journals such as *New Scientist* do not appear on the list because although they are frequently read, they are less frequently cited. The survey does, however, provide some guidelines for libraries with limited resources. The first ten chemical journals, with their rank order in the list of 152 journals, are given in Table 2.1.

Other surveys of journal use have been carried out, for example Stefaniak [26] investigated the primary periodicals that appeared in references selected by running 2,000 profiles against two volumes of *CA Condensates.* The ordering of the key journals in this survey differed from that in Garfield's survey, but *J. Am. Chem. Soc.* still appeared at the top of the list. The *Chemical Abstracts Service Source Index* lists the most frequently cited journals in *CA* [27].

2.2.1.7 *Guides to periodicals*

Many guides to periodicals have been published to enable users to locate primary journals or to check titles and publication data. Some of these guides are listed below and further details can be found in the references [28, 29].

Chemical Abstracts Service Source Index (*CASSI*)
World List of Scientific Periodicals (1900–1960)
Ulrich's International Periodicals Directory
Guides to Scientific Periodicals, M. J. Fowler (Library Association 1966)
Current British Journals (British Library Lending Division and UK Serials Group 3rd edition 1982)
Current Serials Received (published annually by British Library Lending Division)

Table 2.1 – The major chemical journals.

Journal title	Rank of journal in list of 152 key journals
Journal of the American Chemical Society (*JACS*)	1
Journal of Biological Chemistry (*J. Biol. Chem.*)	3
Journal of the Chemical Society (*JCS*)	5
Journal of Chemical Physics (*JCP*)	6
Biochimica et Biophysica Acta (*Biochim. Biophys. Acta*)	8
Biochemical Journal (*Biochem. J.*)	10
Journal of Organic Chemistry (*J. Org. Chem.*)	15
Journal of Physical Chemistry (*JPC*)	23
Chemische Berichte (*Chem. Ber.*)	24
Analytical Chemistry (*Analyt. Chem.*)	30

Keyword Index to Serial Titles (published quarterly by the British Library Lending Division)
Commonwealth Directory of Periodicals (Commonwealth Secretariat 1973)
Union List of Serials in Libraries of the US and Canada
Directory of Japanese Scientific Periodicals (Library of the National Diet, Tokyo, 1974).

2.2.2 Alternatives to the conventional journal

Over the last ten to fifteen years, whilst the majority of journals are still published as print on paper, economic factors and the growth of new technology have resulted in the development of some new forms of primary journal. Conventional journals can be reproduced in microform, although there are very few examples of such journals where there are not corresponding printed editions.

Since 1945, there have been discussions on the use of synopses or abstracts journals, with supplementary microfiche or separates of the full text. The Royal Society of Chemistry publishes the *Journal of Chemical Research*, which started

in January 1977, containing synopses of papers with the supplementary material, chiefly experimental detail, available in both miniprint and microfiche form.

Papers from a central collection can be *repackaged* according to subject interest [30]. In this form of publishing, individual papers may occur in several different selected journals. There are few examples of repackaged journals in chemistry.

The production of supplementary material to published journals is a common practice. In this method of publishing, experimental details, research results such as crystallographic structure data and computer print-out of detailed analyses are not included in the published paper, but deposited, usually in microform, in a repository which may be common to many publications or unique to one.

The United States have had a national depository for many years (the National Auxiliary Publications Service, established in 1937); in Britain, the National Lending Library for Science and Technology started a supplementary publications scheme on the same lines in 1969, which is still operated by the British Library Lending Division; in the USSR, VINITI (the All-Union Institute for Scientific and Technical Information) has run a deposition scheme for scientific manuscripts since 1961 as an alternative to publication in journals, and individual journals or groups of journals with the same publisher may have their own repository systems. The American Chemical Society, for example, includes supplementary material in the microfilm editions of its journals whilst omitting it from the printed editions. This supplementary material is available as microfiche or photocopies.

A further alternative method of publication is the production of *separates* [31]. A number of organizations and university departments publish individual papers in numbered series. These may, or may not, be subsequently included in primary journals. There are not many occurrences of separates publication in chemistry.

The electronic journal is seen as one of the most radical changes in prospect for the primary literature and is discussed in Chapter 3.

2.2.3 Reports and theses

Reports frequently emanate from scientific research carried out under government contract. As much as 85 per cent of the world output of report literature is produced in the USA, but reports are seldom listed in standard bibliographies [32].

Government Reports Announcement and Index (*GRAI*), which is published biweekly by the National Technical Information Service (NTIS) in the USA, announces the American government-sponsored reports. *GRAI* evolved from the *Bibliography of Scientific and Industrial Reports* started in 1946. NTIS also publish *Weekly Government Abstracts* in 26 different series. Microfiche copies of reports can be obtained from NTIS in any one of 500 subject areas as a fortnightly standing order service called *Selected Research in Microfiche* (*SRIM*).

Until 1982, British and foreign reports were recorded in *R&D Abstracts*, published by the Technology Reports Centre in Orpington, Kent, but are now listed in *British Reports, Translations and Theses* published monthly by the British Library Lending Division (BLLD).

Reports and theses provide an earlier presentation of work than a journal article, but the information is often unorganized, provisional or negative. Also, any significant work is almost invariably published subsequently as a journal article in a condensed form. It has been shown that 41 per cent of a sample of British and American chemical papers contained information published in the doctoral thesis of one of its authors [33]. Doctoral theses are, however, a valuable source of original information because each deals with some new aspect of a subject. The preparation of theses involves extensive literature surveys, and a thesis is therefore a useful secondary source of information for locating reviews and other publications on that subject. In spite of their value, there is little use made of theses as an information source, largely because of the difficulty of access to theses [34]. The main guide to theses in Britain is the *ASLIB Index to Theses Accepted for Higher Degrees in the Universities of Great Britain and Ireland.* A number of American and Canadian and some European university doctoral theses are abstracted in *Dissertation Abstracts International*, formerly known as *Dissertation Abstracts* in the period 1952–1969 and previously, from 1938 as *Microfilm Abstracts.* Many theses of chemical interest are also indexed in *CA*.

2.2.4 Patents

Research by several workers has shown that scientists make little use of information from patent documents, although, on average, a patent document is issued in one or other of the major countries every eight seconds of each working day. In the chemical area, less than 10 per cent of the information that appears in patent documents is published elsewhere, and then only very much later [35, 36]. The converse is also true in that less than 10 per cent of the information that appears in the open literature will appear in patents. The two types of literature are thus complementary.

Patent systems have a very long history. In medieval times, monopolies, or exclusive privileges, were granted as favours by ruling monarchs, but the system was abused and in the seventeenth century the Statute of Monopolies laid the groundwork for patent laws which were subsequently introduced in the nineteenth century. The early documents provided at best a scant description of the invention and it was not until the 1700s that the provision of a written description became the normal part of the UK procedure. Documents were not printed for publication until the passing of the 1852 Patent Amendment Act which resulted, among other benefits, in the establishment of the UK patent classification.

The basic principle behind the granting of a patent is that the inventor is given monopoly rights in the use of his invention for a limited period in return for disclosing the nature of that invention in sufficient detail for it to contribute to the general advancement of science and technology. All the major countries developed their own patent systems but, whilst the basic idea was the same, distinct variations developed not only in the length of time of the monopoly granted but also in the nature and concept of the invention. Not everything that might be regarded as an invention is patentable. In the UK, the invention must be novel and involve an inventive step, and it must also be capable of industrial application. Patentable inventions are concerned with technology rather than with the arts or even pure science. Discoveries as such, scientific theories, mathematical methods, aesthetic creations and computer programs *per se* are not patentable.

2.2.4.1 *Patent documents*

A modern patent document comprises three parts. The first is the cover page on which are printed the bibliographic details and usually an abstract. The second is the specification itself which contains the description of the invention and may include background information relating to earlier technology in the field, known as 'prior art', and the reasoning which led to the present invention together with examples, formulae and drawings as appropriate. The third part comprises the claims which set out precisely the scope of the monopoly sought.

Patent documents are very stylized and the use of internationally standardized codes for bibliographic details makes it easy to find information from the texts of most countries, in spite of the fact that patents are written in a legal phraseology known informally as 'patentese', in which words are chosen to impart a precise meaning to the text.

A granted patent gives to the proprietor the right to prevent others from making, using or selling the invention for as long as the patent remains in force. This right applies only in the country or area granting the patent. It is important to realize that whilst the monopoly is restricted, the value of a patent as a piece of literature is timeless and free from any geographical restriction.

A granted patent does not automatically give the proprietor the right to make, use or sell his invention, for in doing so he may infringe the patents of others. A patent tends to be defensive in the same way that publication in the open literature pre-empts or at least diminishes subsequent publication of similar material by others.

2.2.4.2 *Patent systems*

The patent systems that developed in the major countries each involved the submission of a specification to the national patent office, where it remained secret until publication after acceptance or grant. This grant followed an examination of the specification by an examiner at the national patent office to ensure that it satisfied the requirements of the national patent act. The period between the filing of the specification for examination and the date of publication of the granted patent varied from country to country.

An inventor or his agent had to file in each of the countries in which he wished to have protection, and problems could arise if he did not file in every national office simultaneously. This led to the first of several international agreements on patenting policies.

The Paris Convention of 1883, to which over 80 countries now adhere, allowed the inventor one year's grace from the date on which he filed in his national office, during which time he could apply in other countries, in each of which his application would be treated from the point of view of inventiveness and novelty as if it had been filed on the same date as his national application. This date is known as the priority date and the subsequent foreign applications are called convention or equivalent applications. Thus one invention can be protected internationally by a *family* or bundle of patents comprising the first national patent and all its equivalents.

All the national offices were carrying out searches locally on the same invention, and man-hours could be saved by having each invention searched only once. The Institute International Des Brevets (IIB) was set up in the Hague in 1947 to carry out searches on behalf of member countries, within the framework of their patent procedures. Although the UK joined in 1965, it never used the IIB for official searches, but individuals and companies made use of the service.

The scope of the search varies from country to country. In the UK, the search is restricted to UK patents published within the previous fifty years. There have been occasional exceptions, for example where an examiner cited Exodus Chapter 5 against a patent concerned with the manufacture of bricks from fibrous material!

2.2.4.3 *Deferred examination*

The increase in international trade, together with the advances in technology, led to considerable backlogs of patent applications awaiting examination in all national offices. To deal with this problem, many European countries have moved over to a system of deferred examination. In this system, each application is checked for formalities and a search carried out before the application is published, 18 months from the priority date. The search result is sent to the applicant who then has a period of time in which to consider any prior art found in the search and to carry out further work, both of which could lead him to abandon the application. If he decides to continue he must request formal examination. In the Netherlands, where the period for consideration is seven years, only about 40 per cent of all applications proceeded beyond the application stage. By this means, the offices achieved a reduction in their workload and provided quick publication of new ideas. The UK moved over to a form of deferred examination system with the 1977 Patent Act, but with a two-year period from the priority date, in which to request examination.

One disadvantage of the deferred examination system is that it leads to multiple publication of patent documents. For example, in Germany the published unexamined application, the Offenlegeschrift, was followed by the published examined application, the Auslegeschrift, and finally the granted patent, the Patentschrift, was published. The first two documents were photocopies from the typescript which often led to poor copies. To save printing costs the intermediate Auslegeschrift is no longer published and there is a two-stage system similar to that in the UK, where applications are published 18 months from the priority date. The first publication is known as the *A-document* and an example is shown in Fig. 2.2. The prior art found by the examiner is given on the front page. The applicant then has six months to decide whether or not he wishes to go on to substantive examination. If he does and is successful, the granted patent is published as a *B-document*.

---

(12) UK Patent Application (19) GB (11) 2 043 447 A

---

(21) Application No **7905299**
(22) Date of filing **14 Feb 1979**
(43) Application published **8 Oct 1980**
(51) **INT CL³ A01N 25/32**
(52) Domestic classification **A5E** 311 323 504 506 507 E
(56) Documents cited
GB 1484842
GB 1457130
GB 1457129
GB 1457128
GB 1454043
GB 1396942
GB 1396941
(58) Field of search **A5E**
(71) Applicant
Nitrokémia Ipartelepek
8184 Fűzfőgyártelep
(no street)
Hungary
(72) Inventors
Katalin Görög
Erzsébet Dudar
Ivan Gardi
Mária Kocsis
Sándor Gaál
Márta Tasnádi
(74) Agents
T Z Gold & Company

(54) **Herbicidal antidotes**

(57) Selective herbicide compositions comprise at least one herbicide compound optionally selected from the following group: a thiolcarbamate, triazine, chloroacetanilide, carbamide or phenoxyacetic acid in admixture with 0.1 to 50% of an antidote which is a dichloroacetamide derivative of the general formula

$(CH_2)_n$, CH_2, CH_2, X, N – C(=O) – $CHCl_2$, C, R_1, R_2

wherein

X is oxygen or sulphur atom or SO or SO_2,

n is 0 or 1 and

R_1 and R_2 are hydrogen, alkyl, or phenyl substituted with halogen, hydroxyl or nitro; or

R_1 and R_2 together form a butylene, pentylene or hexylene group, optionally substituted with one or two methyl groups,

with the *proviso* that if n = 0, R_1 and R_2 do not stand for hydrogen, alkyl or substituted phenyl, related to the weight of the herbicide compound.

The antidotes may be prepared by reacting a cyclic amine, an N-nitro or N dichloroacetylated hydroxy- or Mercapto-alkylamine, or a Schiff base with e.g. dichloroacetyl chloride, optionally with an alkali metal carbonate, HCl, HBr or p-toluenesulphonic acid as catalyst.

Fig. 2.2 – An 'A-document' from a UK patent application.

There have been two important new developments in the field of international cooperation aimed at reducing still further the demands on patent office manpower. These are the Patent Co-operation Treaty (PCT) and the European Patent Convention (EPC).

2.2.4.4 *The Patent Co-operation Treaty*

The Patent Co-operation Treaty is essentially an agreement between some forty countries throughout the world to accept the search results and preliminary examination made in certain other offices. With the PCT a single international application is filed by the applicant in one language at his national or regional office. This filing has the same effect as if national applications had been filed separately in each of the contracting states designated by the applicant on the application. A search for prior art is made by an International Searching Authority (which includes the patent offices of the USA, Soviet Union, Japan, Sweden, Austria and also the IIB). The application is published 18 months after filing together with the search report. The application and search report are sent to each national patent office of the states designated by the applicant, where it is processed in exactly the same way as a national application. In the UK, such applications issue in the same sequence as the 1977 Act applications, but if the original application was made in the English language then a full specification is not published because the text will already be available in the UK as the published PCT application. Texts in other languages are translated.

2.2.4.5 *The European Patent Convention*

The European Patent Convention (EPC) is available to all countries of Western Europe, Scandinavia, Greece, Turkey and Yugoslavia. Under this convention, the applicant files through his national office to the European Patent Office (EPO) a single application which designates those member states in which he seeks protection. A search is made at the IIB and a search report is published with the application after 18 months. Then, at the request of the applicant, examination is carried out by the EPO. If the application is successful, it is published by the EPO as a granted patent. The document is sent to the patent offices of each of the designated states for acceptance under national law. Whilst there is provision for the UK to demand translation of those European patents not in the English language, this has not been invoked and so there are patents in force in this country which have claims in English, but texts in French or German. This failure by the UK Patent Office to require translations is viewed with concern by many industrialists.

Patents obtained by four different routes can thus be in force in the UK: some from the 1949 Act, newly granted patents from the 1977 Act, PCT based patents numbered in the same series as those from the 1977 Act, and European patents designating the UK.

2.2.4.6 *Patent searching*

Patents can be searched in a variety of ways: name searches, family or equivalents searches, and subject searches, including novelty searches and infringement searches.

Name searches are used to determine the patents held by a given company or individual. Originally these searches were carried out manually but now, wherever possible, the online services are used to search databases such as Derwent's World Patents Index (WPI), *CA*, the various US patent files on DIALOG, Pergamon-InfoLine and SDC, and the QUESTEL INPI-1 and INPI-2 files. Manual searches for older material are still carried out through hard copy name indexes and through INPADOC microfiche.

Family (equivalents) searches are used to determine the international coverage of a particular invention. Such searches are also normally carried out online using WPI or the INPADOC file. The INPADOC file is particularly useful because it gives the publication dates of the family members and covers 50 patent issuing authorities. Older equivalents may be traced using the *CA Patent Concordances*, or the annual name indexes produced by the national patent offices.

Subject searches cause problems which are different from those normally encountered in searching the open literature. In the chemical area, the major problem occurs in searching for generic or ***Markush*** structures, each of which may cover a multitude of specific compounds, most of which will not be exemplified in the text. It is necessary to be able to retrieve each and every specific compound encompassed by the Markush structure since the patent covers each of these. The problems of coding and searching Markush structures are covered in Chapters 5 and 6.

There are a variety of subject searches that can be carried out. The simplest searches are the state-of-the-art surveys which use both the patent and the open literature. It is not essential to have complete coverage for these searches, and interest is normally concentrated on recent advances, so that online files are particularly useful. It is usually sufficient to use WPI, *CA*, UK patent office file lists, US patent files online and possibly some of the more specialized files such as those of the American Petroleum Institute.

The second type of subject search is the novelty search. These searches are made in order to establish that an invention is really new and that it is therefore worth the considerable expense of filing a patent. All of the online files mentioned above are used and additionally, attention is turned to the older patent and open literature.

Infringement searches are the most important searches made. These are used to establish if new products or processes fall within the scope of current, third party patents. These searches have to be exhaustive and much depends on the skill of the searcher who must decide where to search. The search can be restricted to patents issued in the last 20 years, since all others must have expired. Once any relevant patents have been found, equivalents must be sought in all countries under consideration and then a check made to eliminate all those which have been allowed to lapse through non-payment of renewal fees. This leaves the number of patents which are relevant and in force, and these are passed to the patent agents involved for final appraisal. Once the agent has decided which of the patents, if any, would be infringed, these are reported to the searcher who carries out validity studies on each in order to determine whether or not they are soundly based. These searches are also important since if the searcher cannot establish that the patents in force are invalid, then the process may have to be abandoned or a licence sought. Such searches are carried out on patents, all forms of open literature and trade literature. They are time-consuming, but a successful search may save a company considerable sums of money that might otherwise have been spent on licences.

The patent literature is an extremely valuable, but somewhat neglected, primary source of information. The coverage of the patent literature by the online services is discussed further in Chapter 3.

2.2.5 Conferences and personal contact

Conferences are usually organized and attended with four main aims in view:

(1) exchanging ideas and information;
(2) bringing oneself up-to-date;
(3) learning more;
(4) talking and meeting with others in the same field.

These aims can be achieved by presentation of formal papers, discussion sessions, workshop sessions, or poster sessions or by informal and social events. The main disadvantage of a large lecture is that information can be presented only in a linear fashion and little discussion is possible. Discussion and workshop sessions can overcome this limitation, while poster sessions provide the opportunity to combine discussion with a very much greater assimilation of knowledge. Poster sessions are often used in very large congresses, for example the large crystallographic congresses, where it is possible during the course of about three hours to wander round a hundred or more papers and to talk to people whose work is relevant to one's own field of interest. The onus is on the conference delegate to find out as much as possible, rather than just being presented with a lecture to listen to.

One of the main values of a conference is that it provides the opportunity for personal contact, which is an important source of information particularly amongst academic scientists who, for economic reasons, rely heavily on the 'invisible college'. The interactive nature of non-formal communication is a particular advantage over the more formal methods of information transfer.

Conference proceedings are frequently published and details of these appear in publications such as Interdok's *Dictionary of Published Proceedings* from

1964, ISI's monthly *Index to Scientific and Technical Conference Proceedings* produced since 1978, and the British Library Lending Division's monthly *Index of Conference Proceedings Received.* The latter is available as annual cumulations, and an 18-year cumulation covering the years 1964–81 is available on microfiche.

Details of forthcoming international conferences can be found in the *World List of Future International Meetings* published monthly by the Library of Congress, or in Aslib's quarterly *Forthcoming International Scientific and Technical Conferences.*

2.3 SECONDARY PUBLICATION

Secondary publications in the form of abstracts and indexes provide a means of finding relevant journal articles. Title indexes and short indicative abstracts are used to alert scientists to the existence of an article, whereas detailed, informative abstracts are used for comprehensive coverage, as in *CA.* It may take more than a year from the appearance of a journal article before the abstract is included in the abstracts journal.

2.3.1 Abstracts journals

The first major abstracts journal, *Chemisches Zentralblatt*, appeared in 1830, and some British journals, such as the *Journal of the Chemical Society* contained abstracts from 1871. *Chemical Abstracts* started in 1907 and is now the only English-language service abstracting all fields of chemistry and chemical technology. Because of the key role that *CA* plays in covering the chemical literature, chemists rely heavily on its use. At its inception, Chemical Abstracts Service (CAS) was preparing about 12,000 abstracts from 396 journals annually and the number has risen to over 450,000 abstracts annually in 1982 from over 10,000 journals.

The indexing of such an enormous volume of information is of paramount importance. The preparation of the first 10-year accumulative *Subject Index* to *CA* led to the development of a systematic method of naming compounds. Volume indexes are now published semi-annually and collective indexes were produced every ten years until 1956 since when they have changed to five-year cumulations. The *10th Collective Index* (1977–1981) totals more than 125,000 pages in over 80 volumes. A *Numerical Patent Index* appeared in 1911, and an annual *Formula Index* was added to *CA* in 1920. The former was replaced by a more comprehensive *Patent Index* in 1981. An *Index of Ring Systems* was introduced in 1957 and a *Hetero-Atom-In-Context Index* (*HAIC*) was included from 1967.

In 1965 there were two major developments at CAS. A microfilm edition of *CA* was introduced and work was started on the computerization of the preparation of *CA*, such that once the titles and abstracts of papers had been prepared from the primary sources, the printed publications could be produced automatically. In 1969, the whole system was computerized and all chemical compounds indexed were assigned a unique Registry Number. Details of the Registry system are given in Chapter 5.

At the Institute for Scientific Information (ISI) the main chemical information product is *Current Abstracts of Chemistry and Index Chemicus*^{R} (*CAC&IC*^{R}). This weekly printed publication provides abstracts of articles reporting new organic compounds. It was first introduced in 1960 and has indexed more than three million compounds reported new since that date. Each year, the staff of chemists at ISI examine at least 35,000 chemical articles from more than 110 source journals, of which only about 15,000 report new compounds and methods. Each year about 200,000 additional compounds are reported in *CAC&IC*. Entries in this publication are arranged by journal in a graphic format to facilitate scanning, as shown in Fig. 2.3. Each abstract includes structural diagrams and reaction schemes that allow the chemist rapidly to locate new compounds and review them in the context in which they are reported. Included in the abstract are techniques used to identify and purify compounds and an identification of explosive reactions, labelled compounds and new synthetic methods. A use profile alerts the user to potential or tested biological activities and applications. When available, the author's own abstract is included. If there are many ambiguities in a journal article, the authors are contacted for clarification or correction. *CAC&IC* abstracts appear within 45 working days of the publication of the original article in a primary journal.

Each issue of *CAC&IC* contains a set of seven indexes to the articles covered that week. This section is called *Index Chemicus*^{R}. In addition to author, journal, corporate and subject indexes, *CAC&IC* includes indexes to the labelled compounds and biological activities reported in the articles. A molecular formula index and WLN index which guide the user to articles reporting specific compounds are also provided.

Index Chemicus is cumulated quarterly and annually. A *Rotaform Index*^{R} is supplied with these cumulations. The *Rotaform Index* uses each element of a molecular formula (except for carbon, hydrogen, oxygen and nitrogen) as an indexing term for that formula.

Details of other abstracting services in the field of chemistry and related disciplines can be found in reference [37].

2.3.2 Indexes and indexing

Indexes to abstracts are prepared alongside the abstracts, or possibly after the abstracts have been produced. Title indexes can, however, be prepared very much more rapidly than an abstracts publication and can therefore be used for current awareness. Titles of journal articles may be reproduced directly from the contents pages of journals as in the publication of *Current Contents*, produced weekly by ISI. Alternatively, titles can be organized into a KWIC index, as in *Chemical Titles*, produced at CAS.

J. Agr. Food. Chem. 30(1), 1982 MY 916

328944 CONVENIENT PREPARATION OF CHLORDECONE ALCOHOL (KEPONE ALCOHOL) AND ITS DEUTERATED, TRITIATED, AND DECHLORINATED DERIVATIVES.
BLANKE*R V, FARISS M W, SMITH J D, GUZELIAN P S.
VIRGINIA COMM UNIV, MED COLL VIRGINIA, DEPT CHEM, RICHMOND, VA 23298.
J AGR FOOD CHEM 30(1),185-7(1982).

GC MS NMR CCR TLC

Successful synthesis of chlordecone alcohol and its deuterated, tritiated, and dechlorinated analogues is described. Borohydride reduction of chlordecone or of one of its dechlorinated derivatives resulted in the formation of the corresponding alcohol in greater than 60% yield. Characterization of reduction products was established by chromatographic (GLC;TLC) and spectral (IR;NMR;MS) analyses.

USE PROFILE

CARCINOGENIC ACTIVITY
PESTICIDAL ACTIVITY

All unmarked positions in all cpds contain Cl.

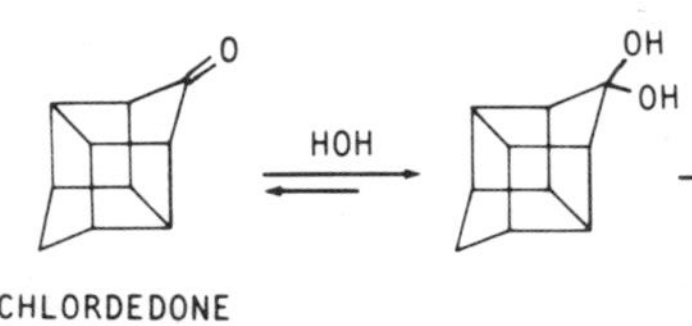

Fig. 2.3 – A sample abstract from *Current Abstracts of Chemistry and Index Chemicus*R.

2.3.2.1 *Subject indexes*

The preparation of subject indexes, such as the Subject Indexes to *CA* requires highly trained staff if the consistency and quality of the index is to be maintained. Manual indexing is in many cases becoming prohibitively expensive, whereas computer power is available at a relatively low cost. KWIC indexing is one method of automatic indexing, but it relies on authors having written informative titles. In preparing subject indexes, the decision as to which keywords are indexed, and whether keywords should be in free text or selected from a thesaurus is a problem for both indexer and searcher. If any new compounds are missed by the indexers at CAS, the retrieval of a particular paper may become impossible. Professor Synge noticed some years ago that five new compounds described in a paper in 1971 [38] were never indexed in *CA* and there was no mention that an optical resolution of an amino acid had been described. This was particularly unfortunate as the amino acid was *p*-hydroxyphenylglycine (the D-isomer of which is used in preparing the popular semisynthetic penicillin 'amoxicillin'), and twenty or more patent specifications appeared over the following years all describing more laborious optical resolutions of the same amino acid. A further problem for the biochemist is the lack of standardization in the indexing of Linnaean binomials in *CA*. The Linnaean system is international and the binomials could be used systematically as keywords by abstractors and indexers.

Automatic indexing from free text is being studied and systems are being developed which depend on the frequency of occurrence of words, co-occurrence with other words and simple algorithms for analysing positions of words [39]. The ASSASSIN system uses a stored word list of important words and a list of stop words. All words new to the system are displayed and the system allows for thesaurus construction [40].

2.3.2.2 *Citation indexing*

A totally different concept of indexing was used by ISI in the *Science Citation Index* (*SCI*), which was started in 1961 [41]. The *SCI* is based on the concept that bibilographic citations reveal subject relationships between papers. The intellectual content of the index is entirely determined by the authors, editors and referees. By tracing back from a given paper, or by tracing forwards (i.e. to all papers that have cited a given key paper) it should be possible to locate the majority of relevant papers. It is also likely that many irrelevant papers will be cited.

It has been noted that American authors tend to cite less than other authors, but the quality of *SCI* could be improved by all authors being more conscientious about the use of citations.

For further information on abstracting and indexing services, see reference [42].

2.4 THE TERTIARY LITERATURE

The tertiary literature is defined in this chapter as evaluated literature based on primary and secondary sources and includes book reviews, handbooks and encyclopedias. When a subject has appeared in a review or handbook, it can be considered to be integrated into the subject literature. Garvey [43] and Bottle [33, 44, 45] have studied the time taken for information to be integrated into the literature. Bottle considered that a subject is fully integrated when it first appears in an undergraduate textbook or in a high school's leaving examination, and he noted that the median period for a topic to reach the schools from the universities is about fourteen years. Important subjects have become integrated much more quickly in recent years than they were in the early part of this century [46].

2.4.1 Reviews

The importance of reviews has been recognized from the earliest times. *Philosophical Transactions* contained review articles in the seventeenth century, but the first review journal as such was *Berlinischer Jahrbuch für die Pharmacie* first published in 1795. The tertiary literature can be written only by those with long working experience in the field, and this is generally reflected in the high quality of reviews. Cuadra [47] has detailed the ideal qualifications of a review writer and states that in addition to having sufficient prestige in the field, the author of a review must be willing to do an enormous amount of reading and evaluation in a short space of time. Several reports have been written on the importance of reviews [48–50], and the SATCOM report [50] in 1969 concluded that critical reviews are essential for effective utilization of scientific and technical information. Because of the large number of references, sometimes over 600, included in a review, it provides an excellent bibliography of the primary literature. Reference lists of reviews are extensively used by chemists for retrospective searches. According to Brunning [51] most chemists read reviews in their own and related fields but they make little use of general scientific reviews. Research chemists need comprehensive reviews on specialized topics with extensive bibliographies. As discussed in Chapter 1, the expert review writer tends to emphasize the most recent advances in a field, as it is not possible to cover every aspect of a subject and thus the review is most appropriate for updating the chemist's knowledge rather than providing an overview of the topic. In the ACSP survey [22] reviews were considered an important source of information for current awareness in spite of the inevitable delays before information appears in a review. Chemists engaged in teaching are less well served by reviews as they require general reviews with a few key bibliographic references.

Review journals can usually be recognized by their titles, which invariably contain such words as *Review*, *Advances in* ..., *Annual Report* ..., or *Progress in* .... Key review journals in chemistry are *Annual Reports on the Progress of Chemistry* and *Chemical Society Reviews*, published by the Royal Society of Chemistry. ISI produce an *Index to Scientific Reviews* and the Royal Society of Chemistry at Nottingham produce the *Chemical Abstracts Review Index* (*CARI*) which is a six-monthly KWIC index to reviews in each volume of *CA* since 1975.

2.4.2 Handbooks and encyclopedias

A large number of handbooks and encyclopedias have been prepared in the field of chemistry and related disciplines, all of which are carefully compiled and edited to provide valuable quick reference sources for a chemist. Some of the main handbooks that have been produced in chemistry include *Lange's Handbook of Chemistry*, the *Rubber Handbook of Chemistry and Physics*, the *Gmelin Handbuch der Anorganischen Chemie* and the *Beilstein Handbuch der Organischen Chemie*. A discussion of the *Beilstein Handbuch* is given below. Encyclopedias, which may be single-volume or multi-volume works, are frequently used by non-chemists who wish to locate a specific chemical fact. Some important reference works in organic chemistry have arisen from companies which produce their own catalogues, and the most widely used of these is the *Merck Index*, which is an encyclopedia of chemicals, drugs and biological activities published by Merck & Co. since 1889. A further major reference work is Heilbron's *Dictionary of Organic Compounds* (*DOC*) and brief details of this and the *Merck Index* are given below. For additional information on these and other reference tools in chemistry, the guides to the chemical literature should be consulted (references [1–6]).

2.4.2.1 *The Beilstein Handbook*

The *Beilstein Handbook* is the largest compilation of information on organic chemistry and has been published for over 100 years. The principal objective of *Beilstein* was, and still is, to provide a comprehensive concentrate of the primary literature as a tool for day-to-day research work. This objective can only be attained if the results published in the primary chemical literature are subjected to a scientifically critical appraisal, i.e. they must be checked for their general soundness and consistency with other findings. In addition, the data produced by this selective processing must reach the user without loss of information content. The great importance of an extensive scientific handbook such as *Beilstein* is due to its unchanging fundamental objective, such that the user is guaranteed a critical, reliable, scientifically processed source of information. A team of 160 workers, including 110 Ph.D. chemists, undertake the preparation of *Beilstein* in Frankfurt.

Beilstein covers all carbon compounds in the scientific literature which have been accurately described, obtained in a state of sufficient purity and whose constitution is known. The description of the compounds covers: the constitution and configuration; natural occurrence and isolation from natural products; preparation, formation and purification; structural and energy parameters of the molecule; physical and chemical properties; characterization and analysis; and salts and addition compounds.

To enable compounds to be located quickly in *Beilstein*, numerous aids and informative brochures have been developed, which can be obtained free of charge from the Beilstein Institute (address in Appendix 2).

In the fourth edition of *Beilstein* which has been in print since 1918, the data published over set periods in the scientific literature on the preparation and properties of all carbon compounds have been compiled. The complete work consists of a *Basic Series* and a *Supplementary Series* covering the periods listed in Table 2.2.

Table 2.2 – The series of the *Beilstein Handbook* (fourth edition).

Series	Abbreviation	Period of literature completely covered	Colour of label on spine
Basic Series	H	up to 1909	green
Supplementary Series I	E I	1910–1919	red
Supplementary Series II	E II	1920–1929	white
Supplementary Series III	E III	1930–1949	blue
Supplementary Series III/IV	E III/IV[a]	1930–1959	blue/black
Supplementary Series IV	E IV	1950–1959	black
Supplementary Series V	E V	1960–1979	(to be published from 1984 on)

[a] Volumes 17–27 of *Supplementary Series III* and *IV*, covering the heterocyclic compounds are combined in a joint issue.

The preparation of the *Fifth Supplementary Series* (E V) which will cover the literature from 1960 to 1979, is now being finalized. The publication of this series, starting in 1984, will mark a significant change in that this, and subsequent series, will be published exclusively in English. As in previous series, the *Fifth Supplementary Series* will eliminate repetition of previous results and multiple publication of the same material, will correct erroneous findings in the light of current knowledge, will recognize and clarify conflicting data, and will relate analogous results.

In 1985, a new *General Index* covering the literature from the inception of organic chemistry through to 1960 will be compiled.

Thus *Beilstein* can give users accurate information, by providing critically evaluated data, and can guarantee that no relevant information is lost.

2.4.2.2 *The Merck Index*

The *Merck Index* has developed from a 170-page alphabetical list of the products of E. Merck & Co. into an internationally recognized 2,000-page reference book containing short descriptions of about 10,000 drugs, pesticides and biologically active compounds. There are over 8,000 printed structure diagrams and a cross-index of about 50,000 synonyms including generic, trivial and systematic names, and trademarks.

The *Merck Index* was produced by traditional manual methods up to the eighth edition in 1968, but with the increasing volume of information and escalating costs of book production, a computer-assisted production method was used for the ninth edition [52]. The use of computers enabled the time taken to produce the book to be reduced from 17 months to 4 months and has led to enormous savings in time spent on index-production, which, for the eighth edition, had taken about two years. In the future, amendments and updates will be made whenever new information becomes available and a fully updated encyclopedia will thus be ready for printing whenever a new edition is needed.

2.4.2.3 *Heilbron's Dictionary of Organic Compounds*

Heilbron's *Dictionary of Organic Compounds* (*DOC*), has since its first publication in 1934, proved to be one of the most useful tools for the organic chemist, chemistry student, biochemist, pharmacist and all others who use organic compounds or require access to information on chemicals. The entries to *DOC* are carefully selected by industrial and academic experts, and only 150,000 of the more important chemicals are included. There are about 50,000 entries in the dictionary, and each entry contains carefully selected information describing the structural and physical properties of a compound and its derivatives, together with alternative names and bibliographic references. There are five Main Work volumes of about 1,200 pages each, and two volumes containing a Name Index, a Molecular Formula Index, a Heteroatom Index and a *CA* Registry Number Index.

The fifth edition of *DOC* was published in October 1982 and was designed and produced as a computer database. A sample entry is shown in Fig. 2.4.

1,1′-Binaphthyl, 8CI **B-10164**
1,1′-Binaphthalene, 9CI
[604-53-5]

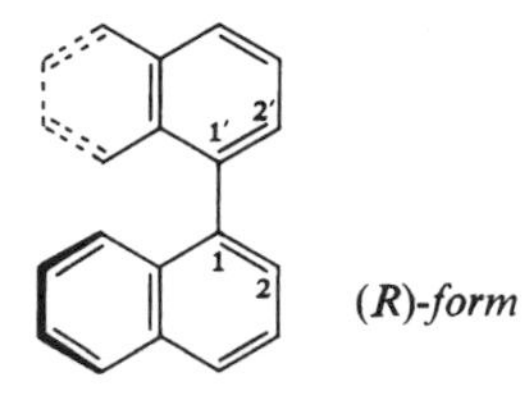

$C_{20}H_{14}$ M 254.331
Atropisomeric compd. with low barrier to interconversion.

(***R***)***-form*** [24161-30-6]
$[\alpha]_D$ −204° (C_6H_6). $t_{1/2}$ ~10h at 25° in several solvents.

(***S***)***-form*** [734-77-0]
$[\alpha]_D$ +222° (C_6H_6).

(±)***-form*** [32507-32-7]
Mp 145°, 159° (dimorph.). Bp >360°, Bp_{12} 290-4°. The cryst. form Mp 159° is a eutectic mixt. of individual (*R*) and (*S*) cryst., the form Mp 145° is a true racemate. The compd. is subject to spontaneous solid-state resolution *via* a phase-transition between the two forms.

Bennett, G.M. *et al*, *J. Chem. Soc.*, 1914, 1057 (*synth*)
Hooker, S.C. *et al*, *J. Am. Chem. Soc.*, 1936, **58**, 1216 (*synth*)
Hodgson, H.H. *et al*, *J. Chem. Soc.*, 1945, 274 (*synth*)
Cooke, A.S. *et al*, *J. Chem. Soc.*, 1963, 2365 (*synth*)
Badar, Y. *et al*, *J. Chem. Soc.*, 1965, 1543 (*ir*)
Morgat, J.L. *et al*, *C.R. Hebd. Seances Acad. Sci*, 1965, **260**, 574 (*synth*)
Browne, P.A. *et al*, *J. Chem. Soc.* (*C*), 1971, 3990 (*ord, abs config*)
Mason, S.F. *et al*, *Tetrahedron*, 1974, **30**, 1671 (*cd*)
Wilson, K.R. *et al*, *J. Am. Chem. Soc.*, 1975, **97**, 1474 (*resoln*)
Kress, R.B. *et al*, *J. Am. Chem. Soc.*, 1980, **102**, 7709 (*cryst struct, polymorphism, bibl*) □

Fig. 2.4 – A sample entry to DOC.

REFERENCES

[1] Subramanyam, K., *Scientific and Technical Information Resources*, New York, Marcel Dekker (1981).
[2] Bottle, R. T., *Use of the Chemical Literature*, 3rd edn., Butterworths (1979).
[3] Maizell, R. E., *How to Find Chemical Information: A Guide for Practicing Chemists, Teachers and Students*, New York, John Wiley (1979).
[4] Antony, A., *Guide to Basic Information Sources in Chemistry*, New York, John Wiley, Halsted Press (1979).

[5] Mellon, M. G., *Chemical Publications, their Nature and Use*, 5th edn., New York, McGraw-Hill (1982).
[6] Skolnik, H., *The Literature Matrix of Chemistry*, New York, Wiley-Interscience (1982).
[7] Borka, H., and Bernier, C. L., *Abstracting Concepts and Methods*, New York, Academic Press (1975).
[8] Borka, H., and Bernier, C. L., *Indexing Concepts and Methods*, New York, Academic Press (1978).
[9] Price, D. J. de S., *Little Science, Big Science*, New York, Columbia University Press (1963).
[10] Bottle, R. T., and Rees, M. K., Liquid Crystal Literature: a Novel Growth Pattern, *J. Inf. Sci.*, **1**, 117–119 (1979).
[11] Harris, F. D., Bibliometrics of Interferon Literature. M.Sc. Thesis, The City University (1981).
[12] Narin, F., and Carpenter, M. P., National Publication and Citation Comparisons, *J. Amer. Soc. Inf. Sci.*, **26**, 80–93 (1975).
[13] Bottle, R. T., Rennie, J. S., Russ, S., and Sardar, Z., Changes in the Communication of Chemical Information I: Some Effects on Growth, *J. Inf. Sci.*, **6**, 103–108 (1983).
[14] Baker, D. Recent Trends in the Growth of the Chemical Literature, *Chem. & Eng. News*, **54**, 23–27 (1976).
[15] Rowland, J. F. B., Economic Position of Some British Primary Journals, *J. Doc.*, **38**, 94–106 (1982).
[16] Sardar, Z., Noise in Written Communication, M.Sc. Thesis, The City University (1975).
[17] Russ, S., The Growth of Scientific Literature, 1900–1970, M.Sc. Thesis, The City University (1979).
[18] Tocatlian, J., Are Titles of Chemical Papers Becoming More Informative?, *J. Amer. Soc. Inf. Sci.*, **21**, 345–350 (1970).
[19] Ziman, J. M., *Public Knowledge: The Social Dimension of Science*, London, Cambridge University Press (1968).
[20] Ravetz, J. R., *Scientific Knowledge and its Social Problems*, Harmondsworth, Penguin (1973).
[21] Garvey, W. D., Lin, N., Nelson, C. E., and Tomita, K., Research Studies in Patterns of Scientific Communication, *Inf. Stor. Retr.*, **8**, 111–122, 159–169, 207–221 (1972).
[22] ACSP, Survey of Information Needs of Physicists and Chemists, *J. Doc.*, **21**, 83 (1965).
[23] The Royal Society, A Study of the Scientific Information System in the United Kingdom, London, Royal Society (1981). (British Library R&D Report No. 5626.)
[24] Ref. [23], p. 12.

[25] Garfield, E., Citation Analysis as a Tool in Journal Evaluation, *Science*, **178**, 471–479 (1972).

[26] Stefaniak, B., Need for Primary Periodicals as Determined by SDI, *J. Chem. Inf. Comput. Sci.*, **21**, 39–42 (1981).

[27] *Chemical Abstracts Service Source Index* (*CASSI*), 1977 and 1978 cumulative supplements.

[28] Ref. [1], pp. 55–59.

[29] Ref. [2], pp. 37–39.

[30] Carroll, K. D., Development of a National Information System for Physics, *Special Libraries*, **61**, 171–179 (1970).

[31] Phelps, R. H., Alternatives to the scientific periodical: a report and bibliography, *Unesco Bulletin for Libraries*, **14**, 61–75 (1960).

[32] Auger, C. P. (ed.), *Use of Report Literature*, Hamden, Conn., Shoe String Press (1975).

[33] Bottle, R. T., Scientists Information Transfer and Literature Characteristics, *J. Doc.*, **29**, 281–294 (1973).

[34] Boyer, C. J., *The Doctoral Dissertation as an Information Source: a Study of Scientific Information Flow*, Metuchen, N. J., Scarecrow Press (1973).

[35] Vcerasnij, R. P., Patent Information and its Problems, *Unesco Bulletin for Libraries*, **23**, 234–239 (1969).

[36] Liebesny, F., Hewitt, J. W., Hunter, P. S., and Hannah, M., The scientific and technical information contained in patent specifications, *The Information Scientist*, **8**, 165–177 (1974).

[37] Ref. [2], ch. 5.

[38] Eagles, J., Laird, W. M., Matai, S., Self, R., and Synge, R. L. M., *N*-Carbamoyl-2-(*p*-hydroxyphenyl) glycine from Leaves of Broad Bean, *Biochem. J.*, **121**, 425 (1971).

[39] Salton, G., and McGill, M. J., *Introduction to Modern Information Retrieval*, New York, MacGraw-Hill (1983).

[40] Clough, C. R., and Kilvington, L. C., ASSASSIN: The Quiet Revolution, *Program*, **12**, 35–41 (1978).

[41] Garfield, E., Science Citation Index – a New Dimension in Indexing, *Science*, **144**, 649 (1964).

[42] Collison, R. L., *Abstracts and Abstracting Services*, Santa Barbara, Ca., American Bibliographic Center (1971).

[43] Garvey, W. D., and Griffith, B. C., Communication and Information Processing within Scientific Disciplines: Empirical findings for Psychology, *Inf. Stor. Ret.*, **8**, 123–136 (1972).

[44] Bottle, R. T., Can Chemists Learn of New Research Efficiently? Talk given to ACS Chemical Literature Group, Wilmington, Del., USA, *Del-Chem Bull.*, **26**, 12 (1970).

[45] Bottle, R. T., Changes in Communication of Chemical Information II: An updated model, *J. Inf. Sci.*, **6**, 109–113 (1983).

[46] Bottle, R. T., Price, C., and Randall, P., Tablets of Stone? How Information Passes from the Frontiers of Research to the A-level course, *Education in Chemistry*, **20**, 211–212 (1983).

[47] Cuadra, C. A., *Annual Review of Information Science and Technology*, New York, Interscience, **1**, 8 (1966).

[48] Royal Society Scientific Information Conference, June 1948, *Report and Papers Submitted*, London, Royal Society (1948), p. 201.

[49] *Proceedings of the International Conference on Scientific Information*, Washington D.C., National Academy of Sciences – National Research Council (1959), p. 649.

[50] *Scientific and Technical Communication: a Pressing National Problem and Recommendations for its Solution*, Washington, D.C., National Academy of Sciences (1969), p. 40.

[51] Brunning, D. A., Review Literature and the Chemist, *Proceedings of the International Conference on Scientific Information*, Washington, D.C., National Academy of Sciences – National Research Council (1959), pp. 545–570.

[52] Windholz, M., Brown, H. D. and Gaspar, T. G., The *Merck Index:* the Merits of Using Computers in Publishing, *J. Chem. Inf. Comput. Sci.*, **18**, 129–133 (1978).

3

Online access to the chemical literature

The use of computers in the preparation of secondary publications, combined with the rapid advances in communications technology, have led to one of the most significant changes in chemical information retrieval: online searching for information. In the 1960s the large secondary information services were unable to cope with the manual task of assembling and indexing the material from the increasing volume of primary literature. Computers were used and thus machine-readable databases were set up as a by-product of the automatic production of printed abstracts and indexes. In the field of information science, the word 'database' is usually restricted to files of bibliographic data and, as a result of a database search, the user must consult the primary source to find the required information unless full text searching is offered. There is also an increasing number of databanks becoming available for online searching. The databanks, or non-bibliographic databases, contain factual or numeric data, which may provide the answers to queries directly, without the need to locate the primary source of the information. Online searching of bibliographic databases is covered in this chapter and the subject of databanks is discussed in Chapter 4.

The retrieval of chemical information online is a vast subject area and one which is constantly changing. Only certain aspects of the subject can be included in this chapter, but further details can be found in references [1–3] and in the periodicals *Online* [4] and *Online Review* [5]. The chapter starts with an introduction to the principles of online searching and shows where online systems differ from conventional methods of searching. One of the major problems that has faced the database producer is that of calculating the appropriate charges for use of the online systems. The criteria used to determine prices are discussed together with the problem of copyright.

The users of online systems have also experienced problems. The large number of databases and systems has led to a corresponding number of different sets of commands, each set being specific to one system, but common command languages have recently been developed to enable several systems to be used with one set of commands. Further problems for the user have been caused by the files not being completely up-to-date although the development of systems based on serial rather than inverted files has shown that the problem of updating files can be overcome. The preparation of the search question is also a complex and time-consuming operation for an inexperienced user, and research is at present being carried out on the use of microcomputer interfaces and the design of new search algorithms which will enable the user to communicate with the system in natural language. These two research areas are outlined to show the direction in which online searching is likely to develop. The second part of this chapter includes brief details of the major chemical database producers and database vendors. For the most recent information on the services provided by these organizations, the relevant News Bulletins and Information Sheets should be consulted. A list of the addresses of all the organizations referred to in this chapter are given in Appendix 2.

3.1 ONLINE SEARCHING

Online searching for information opens up unlimited possibilities for the chemist. He is no longer restricted to searching through a single index for one piece of information at a time. He can, in principle, search online for any of the fields of information on any of the files available to him, and most database producers offer considerably more index terms in the online databases than in the printed index.

The success of online searching depends on the ability of the chemist or information scientist to perform the search in the best possible way, but mainly on the quality, accuracy and consistency of the file to be searched.

The user accesses the system using the terminal, which is connected via a modem or an acoustic coupler to a telephone line, which in turn is linked directly to a local computer or, via a telecommunications network, to a remote computer in another town or country. British Telecom introduced the Packet Switched Service (PSS) in 1981 to enable users in most areas of the UK to make a local call to a PSS node which then links the user to the International Packet Switched Service (IPSS). The latter was the first public intercontinental packet switching data service and was set up in 1978 [6] to allow communication between Britain and North America by connecting with Tymnet and Telenet. The packet switching services transmit information over long distances more efficiently and therefore more cheaply than the conventional telephone networks. The search request is typed at the terminal and transmitted to the computer, which then responds with a preliminary answer. The search request is developed and possibily modified as a result of the response by the computer until the enquirer is satisfied with the results that are returned on the screen. The time taken for the computer to respond to the searcher is almost instantaneous regardless of whether the computer is in the same building or on the other side of the world, and interaction between searcher and computer is accordingly termed 'conversational mode'.

The range of operations that are carried out in an online search include connecting to the computer, logging on to the online system, selecting the file to be searched, entering and combining search terms, printing or displaying search results and logging off. The approach to searching a bibliographic file online is similar to the manual searching of printed indexes in that both methods involve a matching of search terms used by the database producer. In online searching, however, there is a dialogue between searcher and system, multi-faceted searching is possible and the searcher must communicate with the system in the command language appropriate to that system. Owing to the current high level of telecommunication charges, it is necessary before the search can be entered into the computer system to establish a search strategy by deciding which files and systems should be used and then selecting the search terms with guidance from the user manuals and thesauri. There is an unfortunate lack of standardization amongst database producers and thus the size and content of records in a bibliographic file will vary according to the database to be searched. A simple database may contain just the author, title and bibliographic details of the source document, whereas another database may also include a full abstract, subject keywords, publisher, language of document and other details. The records are usually standardized within a given database, although there may be significant differences in the searching facilities for the same database on different host systems. The searchable fields in a database are always defined by the database vendor and it is important to check that the required fields are present before carrying out a search on any particular database.

The searching of chemical bibliographic files is similar to the searching of other scientific bibliographic files, with the exception that there are often a large number of synonyms and ambiguities with chemical names. The searching for chemical names is usually a two-stage process: firstly, the user must search through a chemical dictionary file, where this is provided by the database vendor, to find the preferred or systematic name for the compound. The systematic name, or more commonly a registry number for the compound, is then used to search the main file. It is often necessary when searching for a chemical substance to use a profile of search terms, for example the Registry Number, systematic name and any synonyms, to compensate for the inconsistencies in indexing of compounds on the file.

The problem of ambiguity is overcome in the various chemical structure search systems, in which structures are stored in some form of total structure representation, such as connection table or line notation, as described in Chapter 5. Any terminal used to access the telecommunication networks can be used to input the two-dimensional structure diagram, which is then converted automatically to the unique form for comparison with the file of stored structures for registration or for searching for single compounds. The various methods of substructure searching online are described in Chapter 6.

3.2 PROBLEMS IN THE EVOLUTION OF ONLINE SYSTEMS

3.2.1 Problems for the database producer and database vendor

The rapid development of online systems has given rise to problems for the database producer, database vendor and for the user. The database producer is faced with the problem of controlling the availability of the online systems in terms of costs and legal copyright. The databases are provided as a by-product of the preparation of abstracts and indexes and the database producer has to set appropriate charges for use of the online system compared with the sale of the printed product. The costs of providing secondary services are in two parts: *the first copy cost*, which is the cost of creating the database, and the *run-on cost*, which is the additional cost of producing and distributing each copy of the abstracts journal or computer-readable tape or other storage media. Until now, the majority of database producers have tried to recover first copy costs entirely through the revenue from printed products. This situation is changing, however, because the increase in use of online services combined with the increase in cost of production of the printed abstracts and indexes has led to a substantial decline in the sales of the hard copy [7, 8]. The database producer feels that he is no longer receiving a just reward for the use of online services when they are provided through a database vendor. Only with services such as CAS ONLINE, where the database producer markets the database, is the database producer able to see the relative demand for use of the online service compared with the sales of hard copy, and to adjust the price accordingly.

The Barwise report [9] in 1979 showed that online royalties were being subsidized by the printed products, but suggested that increasing royalties should be charged for online services. Since 1980, online prices have increased more than those for printed abstracts and indexes. Sperr [8] suggests three ways to solve the problem of pricing: the online service price can be set high at the start of the service, or online access could be made available only to subscribers of the printed product, or, alternatively, different data elements could be offered in the online service and the printed product. The database producers are using a variety of methods to encourage the purchase of hard copy in addition to the online services. The pricing policy at the Institute for Scientific Information (ISI) is such that the online service, SCISEARCH, is cheaper to subscribers of the *Science Citation Index* (*SCI*), and Chemical Abstracts Service (CAS) has recently introduced reductions in the cost of the CAS ONLINE service to subscribers of *Chemical Abstracts* (*CA*).

The database vendors use a variety of methods to encourage the use of their online services. An international comparative price guide to online databases is published twice a year in the February and August editions of *Online Review* [10]. Most database vendors offer discounts for increasing usage of their services. For example, BRS allows subscribers to pay for searching as required or subscribers may opt for one of four different subscription levels which depend on the advance commitment made. Other database vendors offer discount to users when their level of use exceeds five hours in any given month.

In addition to the problem of setting appropriate charges for the use of online systems, the database producers and database vendors have the problem of controlling the availability of the information to conform with the copyright laws. In most existing agreements, database vendors specify very limited use of the retrieved material and users are not permitted to transfer the information from the database to a local computer, i.e. to 'download' the information. Database producers are, however, recognizing the downloading problem and in some cases special agreements are offered to users of the databases whereby, for an additional annual fee, they can have unlimited use of the data retrieved. Technological advances have left the laws of copyright far behind and many of the computer systems, particularly the microcomputer systems for data capture and reformatting, probably contravene current, and planned, copyright legislation. In the UK, the Whitford Committee was set up in 1973 to produce a report on modern copyright legislation, but it was 1981 before a Green Paper was produced and the final legislation is likely to be further delayed pending proposals from other EEC countries.

3.2.2 Problems for the user

Online systems have resulted from technological developments, not as a response to users' needs. In fact, when online systems were introduced, some of the improvements which had been made to computer-based systems, such as the use of controlled vocabularly, were abandoned. The techniques of online interrogation were considered to be more important than the accuracy of retrieval. Online searching was, in most cases, introduced into organizations by librarians and information scientists, and the intermediary rather than the chemist, has continued to carry out most online searches to minimize the expense of online searching. Instead of enabling the chemist to have wider access to information, the online systems have taken away his direct access to the information. Most of the database vendors charge for the time spent at the terminal and it is important to keep this time to a minimum. Many organizations will ignore the intangible costs of a chemist manually searching through the literature, but are unwilling to pay for his use of an online service, although the benefits to users, in terms of time saved, are considerable [11].

The cost of an online search is not the only reason why an intermediary is usually responsible for the online searching. A further problem is the number of different commands with which the user needs to be familiar if the search covers several databases. It has been estimated that an online user needs to perform two or three searches per week to maintain proficiency in searching [12], and there is a substantial amount of literature that the searcher must read if he is to keep up to date with changes in the various systems. In organizations where an intermediary is available, it appears that few chemists have shown an interest in online searching.

If the maximum number of relevant references is to be retrieved in an online search, multidatabase searches are essential, but the elimination of duplicate references due to file overlaps gives rise to a major problem for users. It is not feasible to merge the search tapes to provide a unified database [13] and the only solution to the problem is automatically to sort and merge the retrieved references prior to producing the final output [13, 14].

3.3 RECENT DEVELOPMENTS DESIGNED TO HELP THE USER

3.3.1 Common command languages

Although the majority of the online chemical information systems require different sets of commands for access, there have been attempts to develop common command languages to enable several systems to be searched using the same set of commands. A common command language has been developed by the Commission of the European Communities and is offered by some of the database vendors which operate through the European telecommunications network, Euronet. The main commands used in the Euronet Common Command Language are shown in Table 3.1.

Table 3.1 – Commands use in the Euronet Common Command Language.

Command	Function
BASE	Indentifies the database to be searched
HELP	To obtain guidance online
NEWS	To obtain the latest information on system
FIND	To enter a search statement
DELETE	To delete statements or requests
SAVE	To save a search statement for later use
DISPLAY	To display a list of search terms
SHOW	To display or type records online
MORE	To display more data
BACK	To display previous information
PRINT	To print records remotely
STOP	To end a session

3.3.2 Change in file structure

The user of an online system need not be concerned with the structure of the files to be searched online although the method of file organization adopted by a database vendor may have a significant effect on the speed of access to a file or on the currency of the information on the file. If the file is arranged sequentially such that each record follows the previous without any sorting, i.e. a serial file, a search of the database involves a pass through each record in order to test whether or not the record satisfies the search criteria. To speed up the search process for online searching, most database vendors create inverted files whereby the searchable parts of the records are sorted into alphabetical order. Each index term is stored with a list of all the document numbers which contain that term. When the file is searched, the lists are coordinated using a range of Boolean operations to determine the documents which satisfy the search request. Although searching an inverted file provides a rapid response to Boolean search statements, is does entail large computational overheads, because of the extensive sorting operations that have to be carried out in the generation and updating of the indexes. In addition, a huge amount of disc space is needed for the storage of inverted files, and complex software is required to access the various files. Because of the expense of maintaining inverted files, there is now a growing interest in the use of serial, or direct files, which can be updated simply by adding new records to the end of the file. Further advantages in using serial files are that answers to the search question can be provided during the course of the search, as soon as any hits have been found and that the search can be stopped and modified at any stage during the search. In the conventional batch SDI services, the problem with serial files has been the slow response time, but this problem has been overcome in the CAS ONLINE system, in which the total file is divided into several smaller files, each of which is searched by a pair of minicomputers [15]. The minicomputers search the small files simultaneously such that the elapsed time for a search is that taken by a pair of minicomputers to inspect their portion of the file. As the file size increases, more minicomputers are added and the time required to search the complete file remains the same. Examples of fast serial searching machines for text retrieval are described by Hollaar [16] and Maller [17] and these also achieve efficiency in operation by searching in parallel.

3.3.3 Microcomputer interfaces

The rapidly decreasing costs of microcomputer equipment mean that it is now feasible to use a microcomputer to act as a local interface to an external system [18]. Such interfaces can be used to provide a range of facilities for the user of online retrieval systems, including: automatic dialling and logon; the creation, editing and storage of online messages before connection to the host computer; storage of blocks of messages which can be recalled from store and re-used for an online session whenever required; capture of the search output in machine-readable form for editing or reformatting before printing or transmitting to another computer system; and the automatic collection of usage statistics for accounting purposes. It is also possible to use the microcomputer to provide training and help for the new users of a system, to check search statements for syntactic errors, to process natural language queries into a Boolean search statement and to translate a search query so that it can be used on a variety of databases and hosts. The problem of sorting and merging references retrieved from a multidatabase search can also be solved using a microcomputer. Although the use of microcomputer interfaces is not yet widespread, there are several commercial applications. ISI has developed SCI-MATE which enables US users to search DIALOG, BRS, SDC, ISI and MEDLINE via a microcomputer using only one search language. There is a set of menu-driven commands which enable the user to search any of the five systems without previous training. Alternatively, users may search in the original language of the host system, but take advantage of the SCI-MATE facilities for automatically dialling up and logging on to any of the five systems [19]. Many other microcomputer interfaces have been described in the literature [18, 20–23].

3.3.4 Improved search algorithms

Searching a bibliographic file usually involves the use of a Boolean query in which defined search terms are linked by the operators AND, OR and NOT. The items retrieved from the database are those which contain the search terms in the specified logical relationship. Although this method of search is acceptable for most online searching of bibliographic chemical databases, it is less suitable for retrieval from free text [24–27]. Also, the problem of preparing the Boolean search question is normally left to an intermediary. The chemist requires a much more user-friendly system in which he can communicate with the system using a natural language search statement. Research is being carried out on the use of search algorithms which dispense with the need for Boolean statements. The query is expressed as an unstructured list of terms and the retrieved documents are listed in decreasing order of number of terms they have in common with the query. There has been very little use of such search algorithms for online searching because of the complexity and costs of computer processing that are required. Recently, work carried out by Jamieson [28] and Morrissey [29] has involved systems in which a microcomputer processes queries and database output to generate a ranked list of references in order of decreasing similarity. If the user judges the relevance of the retrieved documents, it is possible to modify the query automatically so that further relevant items can be retrieved in a subsequent search and the effectiveness of the search can be greatly increased.

3.4 THE DATABASE PRODUCERS

The US National Library of Medicine was one of the first organizations to create computer-readable files when, in 1964, they made available the database MEDLARS for retrospective batch searching in the USA [30]. The first computer-readable file from CAS was the Chemical–Biological Activities (CBAC) file, produced in 1965 to provide an alerting service to the literature on the biological activities of chemical substances [31]. The creation of machine-readable databases by the secondary information services coincided with the development of long-distance telecommunications networks, such as Tymnet and Telenet in the USA. The online industry was started when two organizations with spare computing capacity, Lockheed [32] and SDC [33] in California, provided the software and necessary computing facilities to enable the databases produced by the information services to be stored and searched interactively via the telecommunications networks. Although there is now a large number of database vendors, or host operators, who provide online access to the chemical literature, there is an increasing tendency for the database producers to offer direct access to their own databases. CAS is the world's key chemical database producer and the CAS ONLINE service, which initially contained information only from the CA Registry Structure File, has recently been extended to allow access to all the online bibliographic information produced by CAS [34].

The remainder of this chapter summarizes the services of the major chemical and chemically related database producers, CAS, ISI, NLM and Derwent, and then gives a brief outline of the various host operators who offer searching facilities for the chemical databases.

3.4.1 Chemical Abstracts Service

Since its inception, Chemical Abstracts Service has been a key provider of information. Computer-readable files were first produced as a result of the automation in the production of *Chemical Abstracts. CA Condensates* contained information typical of most computer-readable bibliographic files:

CA Abstract Number
Names of authors or inventors
Full title of paper or patent
Complete bibliographic citation
Keyword index terms

With the conversion of the semi-annual *CA* Volume Indexes to computer-based production in the early 1970s, CAS introduced the *CA Subject Index Alert* (*CASIA*), a computer file containing the complete General Subject and Chemical Substance Index entries for abstracted documents. *CA Condensates* and *CA Subject Index Alert* files were later combined into CA SEARCH. In the late 1960s, CAS licensed various organizations in the USA and Europe to provide public services from its files. Several of these organizations, notably Lockheed and SDC, began to provide remote online access to the CAS Registry Nomenclature File in the early 1970s. Access to the Chemical Registry Structure File online was first provided through DARC, but, in 1980, CAS themselves offered online access to the Structure File, which then included over five million substances, through the CAS ONLINE service [15], which is described in some detail in Chapters 6 and 7.

Since the end of 1983, in addition to providing a substance search system, CAS ONLINE now also includes the CA File. The latter contains all bibliographic data, including abstracts, from the printed *CA* from 1967 to date in one single file. Registry Number answers from a search of the Registry File can be transferred automatically to the CA File, such that references from 1967 to date can be retrieved for all the Registry Numbers in the answer set. Although CA SEARCH will continue to be offered through the database vendors, CAS have decided not to release the online abstracts file to the vendors because that would probably produce a greater loss of subscriptions to CAS printed services than any other online CAS files. By offering the abstracts directly through CAS ONLINE, CAS can monitor the effect on the sales of the printed abstracts.

In 1984, CAS started the registration of pre-1967 substances and initially they plan to complete the registration of all substances cited in the 6th and 7th Collective Indexes, covering the period 1947–1966.

3.4.2 The Institute for Scientific Information

The Institute for Scientific Information (ISI) is based in Philadelphia and is another major database producer which makes its own databases available directly. One of the major chemical publications produced by ISI is *Current Abstracts of Chemistry and Index Chemicus*^{R} (*CAC&IC*^{R}). The production of *CAC&IC* involves the generation of a machine-readable database called the *Index Chemicus Registry System* (*ICRS*). This database includes all the information in *CAC&IC* except for the authors' abstracts, and covers more than three million organic compounds stored in the form of Wiswesser Line-Formula Notation (WLN). All the information in *CAC&IC* is available online via Télésystèmes-DARC.

A further online chemical database will also be provided: the *Organic Chemistry Citation Index* (*OCCI*). This will be modelled on ISI's new series of disciplinary databases covering the literature of biomedicine, geology and mathematics. In these databases, the literature of each discipline is 'clustered' to reveal the most active research speciality areas. Bibliographies retrieved from those clustered databases represent the most recent papers citing the core, or classic, literature of highly defined speciality areas. These databases can be searched by traditional search terms such as author's names and cited references, or can be searched by 'research fronts': – speciality areas identified through the clustering process.

Since the late 1970s, ISI has been engaged in a long-term endeavour to use clustering to map the entire domain of science. Although the comprehensive work is still some years ahead, a printed prototype of the product has been created: the *ISI Atlas of Science: Biochemistry and Molecular Biology*. In this Atlas, documents identified by clustering have been used to write mini-reviews that describe biochemical specialities. *An Atlas of Chemistry* is planned and

Fig. 3.1 illustrates the relationships between the cited documents that comprise a cluster in organic chemistry. Ultimately it is planned to make the whole *Atlas of Science* available online.

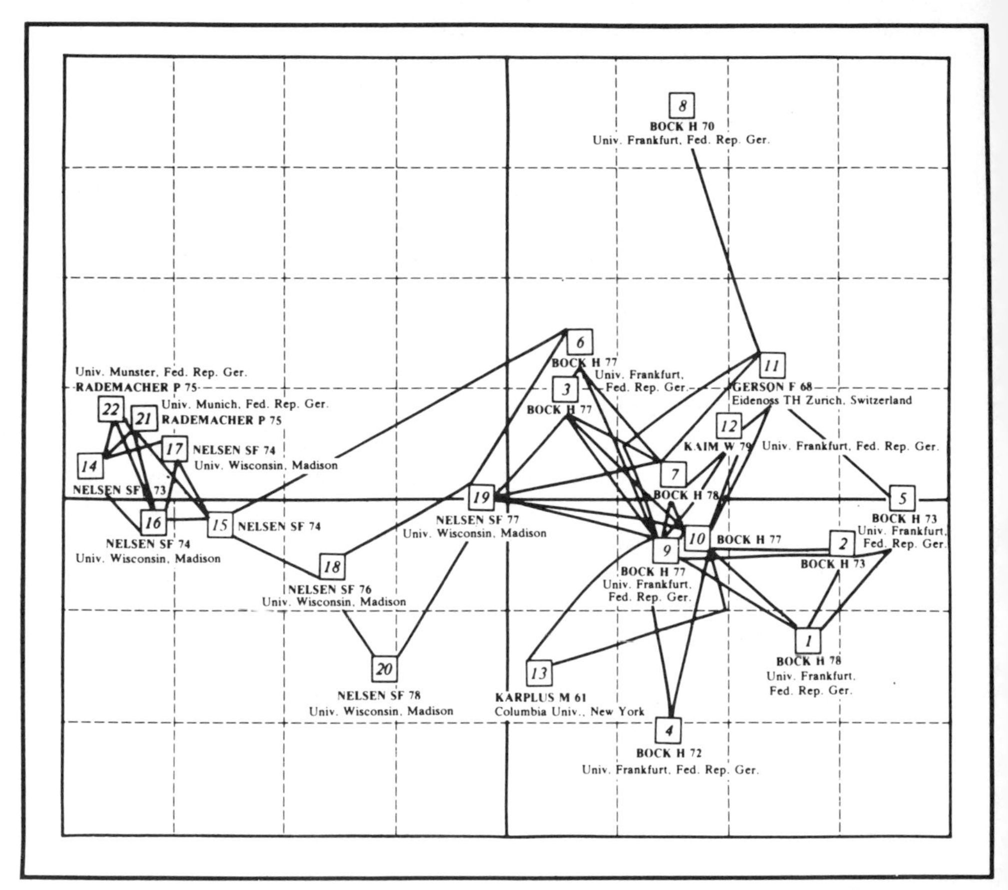

Fig. 3.1 – Sample cluster map from ISI's proposed *Atlas of Chemistry*. Radical ions derived from electron-rich molecules.

3.4.3 The National Library of Medicine

The National Library of Medicine (NLM) was the first organization to create computer-readable files and they were again one of the first organizations to offer an online service, when AIM-TWX (Abridged Index Medicus via the Teletypewriter Exchange Network) was inaugurated in the USA in 1970. The system developed from AIM-TWX was called MEDLINE (MEDLARS Online) and it became operational in 1971 to a limited number of users. The MEDLINE network was greatly enlarged and operates through the Tymshare network with over 50 nodes in the USA and Europe. MEDLINE is now available through BLAISE-LINK, DIALOG, DIMDI and Data-Star in the UK. SDILINE (Selective Dissemination of Information Online) was made available in 1972 as a subset of the MEDLINE file for users who wished to search only the most recent month of the file.

Other databases produced by NLM and which are of interest to the chemist are TOXLINE, CHEMLINE and CANCERLIT. TOXLINE contains citations on toxicology and related topics from six different bibliographic sources. CHEMLINE is an online chemical dictionary and CANCERLIT contains abstracts of citations to the published literature on cancer.

3.4.4 Derwent

The importance of patent information has been stressed in Chapter 2 and it is not surprising that there are a large number of online files that contain patent information. One of the major files is the Derwent World Patents Index (WPI). The Central Patents Index (CPI), which is a part of the WPI, is a comprehensive abstracting, alerting and retrieval service covering all the chemical-type patents emanating from 28 major patent-issuing authorities. The CPI is divided into 12 sections, listed in Table 3.2.

Table 3.2 – The sections in Derwent's Central Patents Index.

Section letter	Subject
A	Polymers; Polymer Technology PLASDOC
B	Pharmaceuticals FARMDOC
C	Agricultural; Vetinary AGDOC
D	Food, Fermentation, Disinfectants, Detergents
E	General Chemicals including Dyes CHEMDOC
F	Textiles, Paper, Cellulose
G	Printing, Coating, Photographic
H	Petroleum
J	Chemical Engineering
K	Nucleonics, Explosives, Protection
L	Refractories, Ceramics, Glass
M	Metallurgy

The complete WPI database, covering over six million patents contained within more than three million patent families is growing at the rate of over

600,000 patents per year. The database, which contains full bibliographic information and special coding, is searchable via SDC, DIALOG and Télésystèmes-Questel. Up to the end of 1980 all the data was contained in a single file, FILE WPI. However, in 1981, it was decided that the Derwent Abstracts would be made both searchable and printable and a new file FILE WPIL (WPI latest) was started.

One of the most important features of the WPI service is the linking together of patents relating to the same invention from different countries, and in some cases the same country, into a single patent *family*. This family is allocated a single Derwent accession number. The first patent is designated as the *basic* patent and all subsequent members of the family are known as *equivalents*.

The types of questions that are directed to a patent file tend to fall into two categories: bibliographic and subject searching. In bibliographic searching, there may be a requirement for a search for the inventor of a patent, or for the company from which a patent issued, or for a specific patent number or for equivalents to a given patent. Subject searching can be carried out by title words, indexing terms, Registry Numbers and, in the case of Derwent's WPI, via codes which represent structural fragments, as described in Chapter 5. Full text searching of abstracts is also available. The online bibliographic search parameters used for searching the WPI are shown in Table 3.3. In addition, various update codes (UP, UPEQ, UPA and UPB) are searchable but not printable.

Table 3.3 – Online bibliographic search parameters in Derwent's WPI.

Print	Information printed (search qualifiers)
AN	ACCESSION NUMBER (AN), INCL. ENTRY YEAR (EY)
TI	DERWENT TITLE (TI), INCL. INDEX TERMS (IT)
PA	PATENT ASSIGNEE (PA); PATENTEE CODE (PC)
IN	INVENTORS (IN)
PN	PATENT NUMBERS(S) (PN), INCL. COUNTRY CODE (CC)
DS	DESIGNATED STATES (DS)
CT	CITATIONS (CT); LANGUAGE INDICATOR
PR	PRIORITY NO. (PR); COUNTRY (PRC); DATE (PRD)
AB	ABSTRACT TEXT (AB)

In addition to the World Patent Index, Derwent also produces the Chemical Reaction Documentation Service (CRDS), and the PESTDOC, VETDOC, BIO-TECHNOLOGY and RINGDOC databases, the latter of which is a specialized database of non-patent literature designed for use in the pharmaceutical industry [35]. Derwent had a 10-year exclusive contract with SDC, which ended in

October 1984, and now WPI is searchable on SDC, DIALOG and Télésystèmes-Questel [36], and there are plans to release other files.

3.5 THE DATABASE VENDORS

Brief details are given below, in alphabetical order, of the database vendors who offer databases of interest to the chemist. No comparisons in terms of costs or services have been given because any such information would rapidly be outdated. The database vendors discussed include BLAISE, BRS, Data-Star, DIALOG, DIMDI, ESA–IRS, Pergamon-InfoLine, SDC, STN and Télésystèmes-Questel. The addresses of all these organizations are given in Appendix 2.

3.5.1 BLAISE online services

BLAISE online services are provided by the Bibliographic Services Division of the British Library, London. The BLAISE-LINE service is based on a computer in the UK and operates the NLM software system Elhill IIIC plus the British Library's EDITOR programs. It provides access to bibliographic records of books, conference proceedings and other material on a complete range of subject areas. For information in the fields of medicine and toxicology, the BLAISE-LINK service provides access to 17 databases mounted at the US National Library of Medicine, notably TOXLINE and MEDLINE. The Registry of Toxic Effects of Chemical Substances, RTECS, produced by the National Institute for Occupational Safety and Health, is also searchable via BLAISE-LINK. In the field of chemistry, access to the CAS Registry Numbers and a database of synonyms for chemical names is provided by the dictionary file CHEMLINE.

3.5.2 BRS

Bibliographic Retrieval Services, BRS, started in 1976 and provides access to over 80 bibliographic and full text databases. It evolved from a system in operation at the State University of New York and runs on mainframes, minicomputers and microcomputers on a software package called BRS/SEARCH. When BRS started operation, it was one of the most economical database vendors. Costs were reduced by basing the charge for the online service on the amount of search time that the customer agreed to purchase in advance: the more search time, the cheaper the rate. All the databases are totally online, but larger files including the Chemical Abstracts files are divided to provide efficient response time. The pricing policies at BRS led to a temporary decrease in prices charged by some of the other suppliers. BRS introduced services in Europe by joining the consortium which operates as Data-Star. A list of the main chemical databases on BRS is given in Table 3.4. Databases in related subject areas include BIOSIS Previews, COMPENDEX, Index to Frost and Sullivan Market Research Reports, Pollution Abstracts and Predicasts databases of business information.

Table 3.4 – The main chemical databases on BRS.

Database	Label	Producer
American Chemical Society Primary Journal Database	CFTX	American Chemical Society
CA SEARCH and *CA Condensates*	CHEM, CHEB	Chemical Abstracts Service
CA SEARCH TRAINING	CAST	Chemical Abstracts Service
DRUGINFO	DRUG, DRSC, HAZE	Drug Information Services, Minneapolis, MN 55455
Kirk–Othmer Encyclopedia of Chemical Technology	KIRK	John Wiley & Sons Inc.

3.5.3 Data-Star

Data-Star started operation in 1980 and was based on Radio Suisse's computer facility in Bern, Switzerland. BIOSIS was the first file to be loaded and the service was opened for European uses via Euronet in February 1981. Since then a large number of databases covering chemistry, engineering, biomedicine and business have been added. The chemist is served mainly by the CAS files which are now offered in one single file. The introduction of 'ZZ' files in 1983 allowed the user to search all the various *CA* files from 1967 to date as if it were one single file. The CHZZ file covers *Chemical Abstracts* and the CNZZ file covers the *CA* Dictionary Files from 1967 to date. The CNZZ file is the first example of a single dictionary file. Other files of interest to the chemist include Chemical Engineering Abstracts and two new databases which were added in 1984: Martindales, which is the British Pharmaceutical Society's Pharmacopoeia, and Kirk–Othmer. The latter provides a full text online version of the 25-volume *Kirk–Othmer Encyclopedia of Chemical Technology*. In January 1984, Data-Star became the UK agent for all BRS services.

The pricing policies for use of Data-Star services are such that users pay a connect-hour charge and royalty charge. Users have the option of a 'no commitment' agreement or four subscription levels based on an estimated annual usage. If users agree to a commitment of more than 480 connect-hours per year, the cost per hour is much less than that of the no commitment user. In addition, academic users are charged at the cheapest rate regardless of commitment. In 1984, it is planned to introduce 'fixed price packages' for use of specific groups of databases where the database producer agrees.

3.5.4 DIALOG Information Services

Lockheed, based in Palo Alto, California, was one of the first of the systems operators to offer online services to chemical databases. Lockheed, now known as DIALOG Information Services, uses the software DIALOG, which evolved from the RECON software developed by Lockheed for the US National Aeronautics and Space Agency [32]. The DIALOG files span more than fifteen years of chemical literature coverage by CAS, in the form of *CA Condensates, CA Subject Index Alext* (*CASIA*) and CA SEARCH, from January 1967 to the present day. DIALOG created a separate file for each of the CA Collective Indexes, starting with the 8th CI, and the files contain a total of nearly five million documents. There is the facility to search any of the individual files or to search comprehensively across all the files using the SEARCH-SAVE feature.

Table 3.5 – Additional databases offered by DIALOG in the field of chemistry.

Database name	Coverage	Database producer
CHEMICAL EXPOSURE	1974–present	Chemical Effects Information Center, Oak Ridge, TN
CHEMICAL INDUSTRY NOTES (CIN)	1974–present	ACS, Columbus, OH
CHEMICAL REGULATIONS AND GUIDELINES SYSTEM (CRGS)	Current	US Interagency Regulatory Liaison Group, CRC Systems Inc., Fairfax, VA
CHEMLAW	Current	Bureau of National Affairs, Washington, DC
CLAIMS COMPOUND REGISTRY	1950–present	IFI/Plenum Data Co. Alexandria, VA
PAPERCHEM	1968–present	Institute of Paper Chemistry, Appleton, WI
SCISEARCH	1970–present	Institute for Scientific Information, Philadelphia, PA
TSCA INITIAL INVENTORY	1979	DIALOG and Environmental Protection Agency, Washington, DC.
SURFACE COATINGS ABSTRACTS	1976–present	Paint Research Association of Great Britain, Middlesex

Information from the CAS Registry System is available in computer-readable form as the CAS Registry Nomenclature File (RNF) from 1972 onwards. DIALOG uses a selection of chemical substances from the RNF to create the CHEMNAME and CHEMSIS dictionary files. Substances are selected for CHEMNAME if they have been cited in *CA* two or more times since 1972, whereas singly-indexed substances are included in the file CHEMSIS. CHEMNAME and CHEMSIS are updated from the CAS RNF quarterly and the files may, therefore, be up to three months out of date. To solve this problem, a new file CHEMSEARCH was created to provide access to all new substances cited in CA SEARCH since the most recent update of CHEMNAME/CHEMSIS. An additional file, CHEMZERO, includes all substances registered with CAS but not cited in CA SEARCH.

All the dictionary files are arranged by citation frequency to be of optimum value to the user, but updating the files is expensive. The dictionary files contain complete *CA* names, molecular formulae, ring data and synonyms, as described in greater detail in Chapter 6. There are no graphical capabilities for substructure searching, but names can be segmented for search and there are element counts, stereochemical descriptions and the new ring data. CA SEARCH is backed by an online thesaurus which contains all the subject terms from the *CA* Index Guide. A further important feature of the DIALOG system is MAPRN which gives the user direct entry to any one of the databases which contain Registry Numbers.

DIALOG offers several other databases of interest to the chemist and these are listed in Table 3.5.

3.5.5 DIMDI

The Deutsches Institut für Medizinische Dokumentation und Information (DIMDI) is a governmental institution within the Federal Ministry of Youth, Family and Health. It provides access to databases in the biosciences and related fields on a cost recovery basis and also covers agricultural information through cooperation with the Zentralstelle für Agrardokumentation und Information (ZADI).

DIMDI uses the GRIPS software (General Relation-based Information Processing System) and offers the NLM and ISI databases of which CHEMLINE, TOXLINE, RTECS and SCISEARCH are of interest to chemists. The large databases on DIMDI are segmented to enable users to obtain faster and cheaper access to the most recent information. Up to a maximum of 5 years of information is maintained in the first segment, after which time the oldest part of the file will be removed to the first backsegment. The backsegments are made available for online searching at specified times, according to a schedule.

3.5.6 ESA-IRS

ESA-IRS is the European Space Agency Information Retrieval Service based in

Frascati, Italy. The first online services were made available in the 1960s using the RECON software developed for NASA. The software has now been further developed and the system known as ESA-QUEST is available via all public networks and directly from Frascati for 20½ hours per day. Access to the CAS tapes is provided, giving full bibliographic information from 1967 to the present day. More than six million references are included covering the 8th to 11th Collective Index periods.

An important feature of the system is that the information is provided online in one integral file, the CHEMABS file, which eliminates the need for cross-file searching. Chemical substances are described by names given by authors, keywords assigned by *CA* indexers, *CA* general subject headings, chemical categories, Registry Numbers and molecular formulae. Concepts are described using the author's own phrasing, descriptions and keyword phrases assigned by *CA* indexers and by general subject headings. ESA-QUEST offers quick response times, straightforward search techniques and some unique search refinements, including the precise control of vocabulary and the ability to search on highly posted substances. The new search command, ZOOM, allows a detailed analysis of the file by giving a frequency analysis of any given terms and enables term associations to be determined.

PASCAL, the database produced by the scientific and technical documentation centre of the CNRS in Paris, is searchable via ESA-IRS [37]. PASCAL is a multidisciplinary database which handles all exact and life sciences in addition to the earth sciences and technology. The database contains over 4.5 million references and is divided into ten subfiles, 25 per cent of which relate to chemistry. Names are given in IUPAC nomenclature and role indicators are added.

3.5.7 INKA

INKA, the Information System in Karlsruhe, was established by the Fachinformationszentrum Energie, Physik, Mathematik GmbH, Karlsruhe, and is financed by the West German Federal Government. The Karlsruhe centre operates the computer and offers 31 bibliographic databases and 15 numeric and factual databanks supplied by various information centres in West Germany. The INKA ONLINE SERVICE is available worldwide via the national and international telecommunications networks.

There are two databases specific to chemistry: DECHEMA and DKI. The DECHEMA database, produced since 1976, contains over 60,000 citations to the chemical engineering literature including chemical process technology, chemical equipment and biotechnology. The database, which is mainly in the German language, is produced by the Deutsche Gesellschaft für Chemisches Apparatewesen e.v. (DECHEMA) in cooperation with the Fachinformationszentrum Chemie GmbH, Berlin, and is supplied by the Dokumentation Maschinenbau (DOMA), Frankfurt. The DKI database is produced by the Deutsches Kunststoff

Institut (DKI), Darmstadt, in cooperation with the Fachinformationszentrum Chemie GmbH, and includes over 120,000 citations to literature on plastics, rubber and fibres, including physical and chemical information on polymers. The database is supplied by the Dokumentation Maschinenbau (DOMA). INKA also offers a number of databases in related subject areas, including metallurgy, physics and geology. All the INKA databases are standardized and can be searched using the DIRS3/CCL Common Command Language. Thesauri are available online, either in the form of an alphabetical list or in hierarchical structure or both.

In addition to the chemical databases, there are two chemical databanks: DETHERM, which is the DECHEMA thermophysical property databank, and C^{13}NMR, which contains numeric values about NMR spectra and is produced by BASF, Ludwigshafen.

3.5.8 Pergamon-InfoLine

Pergamon-InfoLine was established in 1980 to provide access to a large number of public databases in science and technology and is now a major UK service. The key databases of interest to the chemist are CA SEARCH, PATSEARCH, Fine Chemicals Directory (FCD), the INPADOC files, Mass Spectrometry Bulletin (MSB) and the Laboratory Hazards Bulletin. Databases in related subject areas include Chemical Engineering Abstracts (CEA); COMPENDEX; World Textiles; Zinc, Lead and Cadmium (ZLC) Abstracts; PIRA Abstracts; RAPRA Abstracts; World Surface Coatings Abstracts (WSCA) and the database from the Health and Safety Executive, HSELINE.

CA SEARCH is intended for *all* users who wish to search the chemical literature, not just chemists, and is quick and easy to use. The InfoLine command language is designed for ease of use and the new powerful GET command enables users to carry out online analysis of previously created document sets. PATSEARCH, a file of US patents from 1971 to date and PCT applications from their introduction in 1978, contains some previously unavailable patent data that was not put into machine-readable form by the US Patent and Trademarks Office. PATSEARCH is one of the few large bibliographic files that is updated weekly. Pergamon-InfoLine also offer VideoPATSEARCH, a system in which the use of two screens online enables interactive searching to be combined with a display of the patent document. The INPADOC files, produced by the International Patent Documentation Center, Vienna, cover patents issued from over 50 patent issuing authorities throughout the world.

A further important database in the field of chemistry is the FCD produced by Fraser Williams (Scientific Systems) Ltd. The FCD contains some 180,000 catalogue entries from over thirty chemical suppliers and refers to over 60,000 fine chemicals.

Pergamon-InfoLine have developed a system of database clustering by taking a core of major databases in a key area and then adding specialized databases in related areas such that there is degree of overlap.

3.5.9 SDC

SDC Information Services were pioneers in the online industry. They are based at Santa Monica, California, and first offered online access to several large databases in the early 1970s. The software system used at SDC is ORBIT.

SDC offer a wide range of databases in chemical and related areas. The main chemical databases are CA SEARCH and CHEMDEX. CA SEARCH is based on the *CA* Collective Indexes and consists of four files shown in Table 3.6.

Table 3.6 – The CA SEARCH files available on SDC.

File name	*CA* Collective Index no.	*CA* volume numbers
CAS6771	8th CI	67–75
CAS7276	9th CI	76–85
CAS7781	10th CI	86–95
CAS82	11th CI	96–

The file CAS6771 contains 1.3 million citations, CAS7276 and CAS7781 each contain 1.8 million citations and CAS82 is a growing file which is updated every two weeks, when approximately 17,000 records are added. The CHEMDEX database is based on the *CA* Registry files and consists of three files, details of which are given in Fig. 6.3. Molecular formulae are given in CHEMDEX.

The CA SEARCH files are used for searching for bibliographic and subject index information. The *CA* Registry Number provides the link between the CA SEARCH and the CHEMDEX files. Although all the files are purely bibliographic, substructure searching can be carried out to some extent by searching for name fragments in the Name Field of the CHEMDEX files or by searching the ring system description, as described in Chapter 6. The CHEMDEX files can be used, therefore, to find the Registry Number when the *CA* systematic name (8th or 9th CI name) is known, or when part of the name or the ring structure can be specified. The Registry Number is then used for cross file searching in CA SEARCH. SDC plan to add substructure and graphics searching facilities.

SDC also provide many databases in other subjects and they are at present experimenting with database clustering such that areas of overlap in chemistry and related subjects can be established, in a similar way to Pergamon-InfoLine.

3.5.10 STN

The Fachinformationszentrum Energie, Physik, Mathematik GmbH (FIZ), Karlsruhe, are cooperating with CAS to establish and develop the Scientific and

Technical Information Network (STN), which started in 1984. CAS ONLINE is accessible directly from Europe via the STN-Karlsruhe Node, and the PHYSICS BRIEFS database distributed by FIZ in Karlsruhe is available for the first time in North America via the STN Columbus Node. All other databases offered by INKA are being added to STN. The system operates on the *Messenger* software and offers alternative command levels according to whether a novice or an expert is accessing the system.

3.5.11 Télésystèmes-Questel DARC

A major vendor of the CAS files is Télésystèmes-Questel which uses the DARC system. Bibliographic and structural information on over 6 million compounds can be accessed online on the basis of structural characteristics. Text searching on the bibliographic files is carried out using the MISTRAL software and details of the files available for searching on QUESTEL are given in Table 3.7. Enhanced software called Questel-Plus came into operation early in 1984. Structure searching on DARC is described in Chapter 6. The DARC system is available in Europe via Euronet and in the rest of the world via Tymnet and Telenet. In addition to the CAS files, Télésystèmes-Questel are vendors of the PASCAL database, ISI's ICRS and the Derwent databases.

Table 3.7 – The chemical files available on Questel.

Files	Coverage	Number of records	Search language
EURECAS	Since 1965	6,000,000	DARC
POLYCAS	Since 1965	261,000	DARC
MINICAS	Since 1965	60,000	DARC
UPCAS	Last month	25,000	DARC
EUCAS82	Since 1982	403,000	QUESTEL
EUCAS77	1977–1981	2,250,000	QUESTEL
EUCAS72	1972–1976	1,770,000	QUESTEL
EUCAS67	1967–1972	1,310,000	QUESTEL
CANOM	Since 1965	6,000,000	QUESTEL
SPECTRA	Since 1971	40,000	DARC

3.6 THE FUTURE OF ONLINE SEARCHING: FULL TEXT SEARCHING AND DOCUMENT DELIVERY

Online searching has enabled the chemist, with, and occasionally without, the help of an intermediary, to undertake a wide variety of searches of bibliographic databases and to retrieve the required references to the published literature in much less time than a manual search [38]. In addition, the list of references can be printed out so as to alleviate the problem of copying endless references from abstracts journals. However, the problem for the chemist does not end with the receipt of a list of references since there is still a considerable amount of work to be done in searching for source documents in the library. A *databank* search, in which the required information is obtained directly, overcomes this problem and thus offers substantial benefits to the chemist, as discussed in the next chapter. The introduction of full-text searching and document delivery systems opens up exciting new prospects for the chemist and for the future of online retrieval from bibliographic databases.

The term 'electronic journal' has been loosely applied to various theoretical or experimental systems in which a computer is used to aid some or all of the normal procedures of journal production [39]. The term may also be applied to systems in which the full text of journals is available in electronic form for online access.

One of the first electronic journals to be offered as an online database through a database vendor was in the field of biomedicine when in 1983, the International Research Communication System (IRCS) medical science group of journals was made available on BRS and was published both as an online database and in traditional printed journal form [40]. Since the beginning of 1984 IRCS has also been available on DIMDI.

During the last few years, the American Chemical Society (ACS) has been exploring the feasibility of providing the full text of journals in electronic form [41]. In 1980, ACS tested a small file of computer-readable articles that were mounted as a private database at BRS. In 1981, the file was expanded to contain the full text of 16 ACS journals and was updated every two weeks. About three hundred volunteers evaluated the usefulness of the database by accessing online the full text of articles except for tables and some graphic materials. Participants were provided with users' manuals and with search examples. They were given 1.5 hours of free connect time, including telecommunications costs, so that they could become familiar with access protocols and the nature of the file. Those who wished to continue use of the file were charged a nominal fee.

Reactions to and opinions about access to full text were obtained by means of online query response to questions posed after each search session and by follow-up survey questionnaires. Face-to-face discussions were held with several groups of participants and telephone contacts were made with a number of others. The reaction to the availability of the ACS journals online was sufficiently favourable to warrant further investigation. The second phase was begun in May 1982 with the selection of about 250 participants who volunteered to help to evaluate the usefulness of the database of 25,000 articles from 18 ACS primary journals published since 1981. An initial 1.5 hours of free time was again provided and at the end of the experiment, information will be sought concerning the improvements and enhancements in searching and display of retrieved data. In addition, attention will be given to education, training and marketing considerations.

A further type of electronic journal is exemplified in the work of the British Library-sponsored research project BLEND (Birmingham and Loughborough Electronic Network Development) [42]. In this project, authors enter the text of papers into the system and the process of editing, refereeing and disseminating the text will be via computer terminals. Further work on online communications will be studied during the project, such as cooperative writing of papers, the preparation of News Bulletins and the compilation of bibliographies.

In addition to the provision of the full text of journals for online searching, document delivery systems are being developed, in which the printed copy of a journal or book can be ordered online as a result of an online search. Ultimately, the hard copy could be printed on demand, which would lead to considerable savings on the part of the publishers, as it would reduce the cost of warehousing and would also avoid books being out of print. The publishers Blackwells in London are developing both online book ordering and a document delivery service whereby books would be printed when required, but a study by Pira for the Publishers Association of the UK and the Commission of the European Communities (CEC) [43] has shown that it may not be feasible for publishers to set up document delivery services independently because of the investment and risks involved.

The success of a document delivery service is likely to depend on its comprehensiveness of coverage, as users would not be satisfied if they received only some of the requested documents immediately but had to wait several weeks for the others. The CEC is currently financing a number of major experiments to demonstrate the viability of an electronic document delivery service or an electronic publishing service [44], and has commissioned various reports including the Franklin, Steria and ARTEMIS reports. The ARTEMIS report [45] concluded that electronic delivery of primary documents is feasible, but that market forces alone would not be sufficient to bring about electronic document delivery systems. The EEC has also provided financial assistance for the ADONIS project [46] which was a joint venture by a number of scientific and medical publishers to set up an automatic document delivery system throughout the world, but the technology has proved extremely expensive. In the UK, the Hermes project [47] aims to show the feasibility of using Teletex technology for document delivery.

Automatic document delivery is still in its infancy and considerable work has to be done to establish economic systems that are acceptable to both publisher and user. If Lancaster's forecast of a possible paperless information system in the year 2000 [48] is to be realized, online systems must become very much more widely used, particularly by the end-users themselves. The availability of microcomputer interfaces and the introduction of more user-friendly languages to assist the chemist should enable him to appreciate the potential of online searching and once again to accept the responsibility for his own literature searching.

REFERENCES

[1] Hall, J., M., Brown, M. J., *Online Bibliographic Databases*, 3rd edn, London, Aslib (1983).
[2] Houghton, B., and Convey, J., *Online Information Retrieval Systems*, 2nd edn, London, Bingley (1984).
[3] Henry, W. M., Leigh, J. A., Tedd, L. A., and Williams, P. W., *Online Searching: An Introduction*, London, Butterworths (1980).
[4] *Online* published every two months by Online Inc., 11 Tannery Lane, Weston, CT 06883, USA.
[5] *Online Review* published every two months by Learned Information Ltd., Besselsleigh Road, Abingdon, Oxford OX13 6LG, England, *or* Learned Information, Inc., 143 Old Marlton Pike, Medford, NJ 08055, USA.
[6] Casey, M., Packet Switched Data Networks: An International Review, *Information Technology: Research and Development*, **1**, 217–244 (1982).
[7] Lancaster, F. W., and Goldhor, H., The Impact of Online Services on Subscriptions to Printed Publications, *Online Review*, **5**, 301–311 (1981).
[8] Sperr, I. L., Online Searching and the Print Product: Impact or Interaction?, *Online Review*, **7**, 413–420 (1983).
[9] Barwise, T. P., *Online Searching: The Impact on User Charges of the Extended Use of Online Information Services*, Paris, International Council of Scientific Unions Abstracting Board (ICSU AB) (1979).
[10] International Comparative Price Guide to Databases Online, *Online Review*, **8**, 105–112 (1984).
[11] Markee, K. M., Economies of Online Retrieval, *Online Review*, **5**, 439–444 (1981).
[12] Ref. [3], p. 96.
[13] Martyn, J., Unification of the Results of Online Searches of Several Databases, *Aslib Proc.*, **34**, 358–363 (1982).
[14] Onorato, E. S., and Bianchi, G., Automatic Identification of Duplicates after Multi-database Online Searching, *Online Review*, **5**, 445–451 (1981).
[15] Farmer, N. A., and O'Hara, M. P., A New Source of Substance Information from Chemical Abstracts Service, *Database*, **3** (4) 10–25 (1980).
[16] Hollaar, L. A., Unconventional Computer Architectures for Information Retrieval, *Ann. Rev. Inf. Sci. Technol.*, **14**, 129–151 (1979).
[17] Maller, V. A. J., The Content Addressable File Store – CAFS, *ICL Tech. J.*, **2**, 265–279 (1979).

[18] Williams, P. W., and Goldsmith, G., Information Retrieval on Mini- and Micro-computers, *Ann. Rev. Inf. Sci. Technol.*, **16**, 85–112 (1981).

[19] Stout, C., and Marcinko, T., SCI-MATE^{TM}: A Menu-Driven Universal Online Searcher and Personal Data Manager, *ONLINE*, **7**(5), 112–116 (1983).

[20] Toliver, D. E., Ol'Sam: an Intelligent Front End for Bibliographic Information Retrieval, *Information Technology and Libraries*, **1**, 317–326 (1982).

[21] Nicholson, D. M., and Petrie, J. H., Using a General Purpose Micro for Online Searching, *Aslib Proc.*, **35**, 354–357 (1983).

[22] Meadow, C. T., A Computer Intermediary for Interactive Database Searching, *J. Am. Soc. Inf. Sci.*, **33**, 325–332 (1982).

[23] Pollit, A. S., An Expert System as an Online Search Intermediary, *Proc. 5th Int. Online Inf. Meeting*, Oxford, Learned Information, (1981), pp. 25–32.

[24] Verhoeff, J., Goffman, W., and Belzer, J., Inefficiency of the Use of Boolean Functions for Information Retrieval, *Comm. ACM*, **4**, 557–558, 594 (1961).

[25] Salton, G., Dynamic Document Processing, *Comm. ACM*, **15**, 658–668 (1972).

[26] Robertson, S. E., The Probability Ranking Principle in Information Retrieval, *J. Doc.*, **33**, 294–304 (1977).

[27] Stibic, V., Influence of Unlimited Ranking on Practical Online Search Strategy, *Online Review*, **4**, 273–279 (1980).

[28] Jamieson, S. H., The Economic Implementation of Experimental Retrieval Techniques on a Very Large Scale Using an Intelligent Terminal, *ACM SIGIR Forum*, **14**, 45–51 (1979).

[29] Morrissey, J., An Intelligent Terminal for Implementing Relevance Feedback on Large Operational Retrieval Systems, *Lecture Notes in Computer Science*, **146**, 38–50 (1983).

[30] Harley, A. J., An Introduction to Mechanical Information Retrieval, *Aslib Proc.*, **30**, 420–425 (1978).

[31] CAS Report, The First 75 years of Chemical Abstracts Service, Chemical Abstracts Service, Columbus, Ohio (1982).

[32] Summit, R. K., DIALOG Interactive Information Retrieval System, in Kent, Allen, Lancour, Harold (eds.), *Encyclopedia of Library and Information Science*, 7, 161–169, New York, Marcel Dekker (1972).

[33] Cuadra, C. A., SDC Experiences with Large Databases, *J. Chem. Inf. Comp. Sci.*, **15**, 48–51 (1975).

[34] *CAS ONLINE News*, **3**, 1 (1983).

[35] Bawden, D., and Devon, T. K., RINGDOC The Database of Pharmaceutical Literature, *Database*, **3**(3), 29–39 (1980).

[36] DIALOG and Derwent Agree, *Monitor*, **36**, 5–6 (1984).

[37] Pelissier, D., PASCAL Database File Description and Online Access on ESA-IRS, *Online Rev.*, **4**, 13–31 (1980).

[38] Johnston, S. M., Choosing between Manual and Online Searching – Practical Experience in the Ministry of Agriculture, Fisheries and Food, *Aslib Proc.*, **30**, 383–393 (1978).

[39] Turoff, M., and Hiltz, S. R., The Electronic Journal: A Progress Report, *J. Am. Soc. Inf. Sci.*, **33**, 195–202 (1982).

[40] Buckingham, M. C. S., Franklin, J., and Westwater, J., IRCS Online: Experiences with the First Electronic Biomedical Journal, *Proc. 7th Int. Online Inf. Meeting*, Oxford, Learned Information (1983).

[41] Cohen, S. M., Schermer, C. A., and Garson, L. R., Experimental Program for Online Access to ACS Primary Documents, *J. Chem. Inf. Comp. Sci.*, **20**, 247–252 (1980).

[42] Shackel, B., The BLEND System: Programme for the Study of some 'Electronic Journals', *The Computer J.*, **25**, 161–168 (1982).

[43] Gates, Y., User Needs and Technology Options for Electronic Document Delivery, *Aslib Proc.*, **35**, 195–203 (1983).

[44] Vernimb, C., and Mastroddi, F., The CEC Experiments on Electronic Document Delivery and Electronic Publishing, *Proc. 7th Int. Online Inf. Meeting*, Oxford, Learned Information, (1983), pp. 119–130.

[45] Norman, A., Electronic Document Delivery. The ARTEMIS Concept for Document Digitalization and Teletransmission, Oxford, Learned Information (1981).

[46] Stern, B. T., Adonis, Paper presented at the 6th International Online Information Meeting, London, December 1982.

[47] Yates, D. M., Project Hermes, *Aslib Proc.*, **35**, 177–182 (1983).

[48] Lancaster, F. W., *Towards Paperless Information Systems*, New York, Academic Press (1978).

4

Databanks

4.1 INTRODUCTION

The terms databank, non-bibliographic database and numeric database are similar in meaning, the latter term being the most specific. This chapter discusses databanks in chemistry in the broad sense of machine-readable collections of factual information. Data, as usually understood in chemistry, are numerical representations of the magnitudes of various quantities. The term can also include basic qualitative data such as specific, but non-numerical scientific facts, for example, the structures of molecules. The term databank has a wider meaning than that of files of data, as just defined. It is used to mean a collection of numeric, textual or factual information which minimizes the need to access original sources. Many handbooks (hardcopy or microform) can be considered as databanks but in this chapter only those files that are available in machine-readable form are discussed. The growing importance of online retrieval of non-bibliographic information and the salient features of databanks have been reviewed recently [1].

4.1.1 Importance of data

Ideas and new hypotheses frequently stem from examination of data. Data are needed to compare and test the validity of work. In order to ensure maximum benefit of knowledge accrued so far, data needs to be expressed in standard forms to make it exchangeable. A scientist going to the literature is often looking for data. Manual handbooks in existence are of enormous value as is illustrated by their frequent presence in the laboratory.

The volume of data generated makes it impossible for individuals to set up their own collection. Chemists have collaborated in areas of common interest to develop jointly a collection on a particular topic. This often brings together the academic and industrial chemist.

4.1.2 Handling of data

The need for special provision for handling of scientific data has been recognized and an international organization, CODATA, the Committee on Data for Science and Technology, has been created. CODATA was established by the International Council of Scientific Unions (ICSU) in 1966 to promote and encourage throughout the world the production and distribution of collections of critically selected numerical values of properties of substances of importance to science and technology. Their 1980 overview and source book [2] illustrates the scope of their work.

As described in Chapter 2 the article in a primary journal remains the main means of reporting scientific results. The growth in the literature being published has repercussions for the publication of data. The volume of material presented for publication strains the capacity of the journals. The strain leads editors to impose restrictions on the number of pages which the author can devote to comprehensive presentation of his data – frequently only summary tables, examples of graphs etc. can be included. The full exposition of measurements, calibrations etc. must often be sacrificed. As a result only a fraction of the data is immediately accessible to the reader and the data that are presented cannot, in many cases, be adequately interpreted, analysed or compared to other results. Some databanks are now more comprehensive than the primary literature; for example, the Cambridge Structural Database has acted as a depository of unpublished atomic coordinates from a number of journals since 1977. The sheer volume of coordinate data and pressure on journal space is likely to increase this activity in the next few years. Thus the need for databanks has arisen.

In the broader definition of databanks as covered by this chapter there are included 'handbooks online'. The advantage of creating an online version of a handbook is the flexibility of access by terms and data values not searchable in the hardcopy version, e.g. by a group of properties.

4.1.3 Numerical data indexing

Numerical data are produced at great expense, and often in great volume. Most of these data will be analysed only for the single purpose for which they were originally collected even though they may have potential value for other uses. The difficulty of finding specific data, even that reported in the literature, is one of the major reasons for the lack of multiple uses of data. Numerical data indexing [3] is a method that attempts to overcome this limited use of reported data. One of the advantages of numerical data indexing is to make it easy for a researcher to find several papers containing numerical data of the same event so that judgement can be exercised on the accuracy of a value.

Numerical data indexing is an extension of subject indexing and involves the use of data flags and data tags. The data flag is an indication of the presence of numerical data in an article, while the data tag is a more descriptive indexing activity which characterizes the numerical data in greater depth. This type of indexing is designed as additional indexing to the traditional bibliographic handling of scientific information.

In several of the databanks discussed later in this chapter attempts are made to check the data and in some cases to evaluate its quality prior to storage in the databank. Numerical data indexing is not as valuable as the work carried out by data analysis centres, such as those producing the Mass Spectrometry Databank and the Cambridge Structural Database. The work of these centres is described later in the chapter (sections 4.3.3 and 4.4.2).

4.2 GENERAL ASPECTS OF DATABANKS

4.2.1 Types of databanks

Databanks in chemistry can be classified into the following broad categories:

Environment
Legislation
Toxicology and biological properties
Spectroscopy
Chemical and physical properties
Special applications of chemicals, e.g. as drugs or fine chemicals, and data related to these uses

The impetus for compiling databanks in these different categories varies. National and international legislation determines much of what is to be collated into databanks in the environmental, toxicological and legislative fields. The impetus behind the compilation of databanks of spectra and property data comes from the specialists who are generating and using the data. With the special applications the impetus generally comes from the industry involved together with the private sector publisher or vendor.

Examples of the effect of legislation on the creation of databanks are certain components of the NIH-EPA Chemical Information System. For example, the TSCAPP file (see section 4.4.1) arose from the data collected under the Toxic Substances Control Act in the USA. The European Inventory of Existing Commercial Substances (EINECS) has been compiled as a result of an EEC directive (see section 4.4.3).

The Mass Spectrometry Databank was started by a group of mass spectroscopists and the story of its development can be regarded as classical (see section 4.3.3). Similarly the Cambridge Structural Database (CSD) was created by crystallographers for crystallographers, though it now has an increasing audience of chemists, pharmacologists and molecular biologists (see section 4.4.2).

The 'Hansch databank' which is a file of partition coefficients is produced by Pomona College in California as part of their Medicinal Chemistry project. This is an example of a databank being compiled by a specialist group for their own needs; copies of the databank on tape are made available to other organizations. Another databank produced at one centre, initially for internal use, is the carbon-13 nuclear magnetic resonance databank developed at BASF AG. This is now publicly available online via INKA.

The Fine Chemicals Directory (FCD) (formerly known as CAOCI) is a directory of the suppliers of compounds in research quantities and is an example of a databank compiled jointly by companies in the industry that required it [4].

A large volume of spectral data is only available in printed form. A recent guide [5] to the published collections of spectral data held by the Science Reference Library gives an indication of the volume and nature of data available. The status of infra-red spectroscopy databanks has been reviewed [6]. Some collections of spectra are available only to those who purchase particular intruments.

4.2.2 Special data elements and their particular problems

Databanks in chemistry generally contain several fundamentally different types of data element. The main types of data element are text, numerical values, coordinates, codes and structure diagrams.

Each of these poses its own particular problems.

4.2.2.1 *Textual material*

Examples of this type of information are names in chemical dictionaries and descriptive fields in files such as the Oil and Hazardous Materials Technical Assistance Data System (OHMTADS), one of the components in the NIH-EPA Chemical Information System. This file holds up to 126 different fields giving information about a particular material. Some of these fields cover fire precautions and sources of materials for use against hazards in an emergency. In this type of field it is very probable that the information given may become out of date. For example, the type of breathing apparatus recommended for 1984 may well be superseded in 1990. Keeping such text fields up-to-date is a task not met in bibliographic files, which are not edited and validated. In databanks the user expects to be presented with up-to-date and accurate facts.

4.2.2.2 *Numeric values, coordinates and codes*

Numeric values pose little problem in storage and updating. The problem is in judging the quality of the figures submitted and deciding whether the databank should contain all values submitted for a given answer with details of experimental methods where available (as is done in the Register of Toxic Effects of Chemicals, RTECS) or whether a value judgement or measurement should be made if two values of the same parameter are submitted. A value judgement by the databank producer is probably dangerous since the searcher is prevented then from making his own judgement on which may be the preferred value. A true verification of the data is extremely costly. A pragmatic solution is to use quality tests so that the value likely to be of the highest accuracy is chosen for input to the databank. For example a quality index may be used for choosing between different mass spectra for the same compound (see section 4.2.3).

Numeric data necessitates special features in the search system to allow for the 'range', 'greater than', 'less than', 'equal to' type of searching. Furthermore, it is no great advance to be able to retrieve a specific value of data by machine rather than via a handbook or listing. The computational facilities of a mechanized system allow the manipulation of the data retrieved, thus providing a considerable advantage over retrieval from a handbook or listing.

Spatial data represented by coordinates do not pose any particular problem in storage and updating but may necessitate special equipment for display of results. If 3D coordinates are available, graphics software and hardware can be used to display and manipulate the chemical structure representation.

Codes frequently used in databanks are molecular formulae, fragment codes, hazardous material codes, taxonomic codes etc. Special search techniques are required and standardization of expression between databanks is useful.

4.2.2.3 *Structure diagrams*

Although the techniques of searching and output described in Chapter 6 are employed in connection with databanks, special problems exist when printed products are required from a databank. Structure diagrams in the Dictionary of Organic Compounds, Fifth Edition (DOC 5) and in the Cambridge Structural Database (CSD) are required to be represented to a higher standard in a printed product than that required for display as computer output to a search. The DOC 5 computer 'databank' has been used for production of the dictionary and it is not available publicly for searching. However, the techniques for handling the structure diagrams are of interest.

DOC 5 contains some 25,000 structure diagrams not including linear structural formulae, such as (H3C)2CHCH2CH2OH, which are readily handled as normal typesetting. The structures are the single most important data element of the Dictionary. Each one had to be re-made for the new edition and it was considered vital that they should be drawn to the highest possible standard. Stereochemistry needed to be clearly shown (including absolute configuration where known) and the drawing conventions used needed to conform to best current practice. In addition, wherever possible, closely related compounds should have structure diagrams drawn to correspond in orientation so that the user can make rapid accurate comparisons.

The traditional method for incorporating structure diagrams in a work such as DOC 5 is for them to be manually inserted into the page layout prior to blockmaking. This entails either making up the structure from typeset elements such as benzene rings, or for more complex structures, submitting to the printers an artist-prepared diagram from which a block is made. For DOC 5, which was to be printed by phototypesetting, this would involve preparing a photographic print of each structure which would then be inserted by 'stripping-in' into the pages. Although this method would probably be the cheapest, its disadvantages for a databank publication would be that the databank would contain no record whatsoever of the structures and for future editions the 'stripping-in' process would have to be repeated. 'Stripping-in' is a time-consuming process and would add 3–4 months to the production time.

At the other end of the scale, the most technologically advanced method of storing structures is in algebraic form, the structures being input at special terminals. Such technology is now in use by Chemical Abstracts in their Online Structure Input System (OLSIS). However, such technology was rejected for DOC 5: firstly because it would be excessively expensive, requiring investment in hardware and software that could not be justified; secondly because application of an advanced technology would have to be achieved over a short time-scale with no alternative in the case of failure; thirdly, because available systems did not produce complex structures to the aesthetic standard necessary for a high-quality printed book.

A third, intermediate technology was adopted for DOC 5. Structure diagrams are prepared by an artist and then raster-scanned to convert them to a digital record. This is stored on magnetic tape which is kept separate from the text tape which contains 'markers' indicating the position of each structure in the text. This method is not ideal; it has the disadvantages that the digital scanning is a slow process and that the digitized structures are not amenable to potential future uses, especially online substructure searching. However, the printed results are first class. If DOC 5 becomes heavily used in online applications, algebraic manipulation of structures may become of importance instead of, or alongside, digitization.

Another databank from which high quality printed structures are required is the Cambridge Structural Database. Until recently the only segment of CSD (described fully in section 4.4.2) that could not be recognizably reproduced on output was the chemical connectivity which could not be reproduced as the two-dimensional chemical diagram from which it was derived. Algorithmic generation of chemical diagrams from connection tables is possible, but no method could be found which was suitable for application to the very broad spectrum of compounds present in CSD, especially bearing in mind the high proportion (37 per cent) of organometallics and metal complexes.

The solution adopted is the reverse of the algorithmic approach: the connection table is generated from a digitized diagram, which can be readily drawn from the 2D x,y-coordinates of each atom also obtained in the procedure. The first stage is direct digitization of rough hand-drawn copy (as shown in Fig. 4.1). The molecular formula is also input as a text string. A series of BASIC programs permits editing and stylizing of the initial diagram (as shown in Fig. 4.2). The following stages are included:

(a) The molecular formula generated from the diagram is compared with the input text.
(b) Necessary chemical edits are effected. Here the N=C double bond (arrowed) replaces the incorrect C–C bond.

(c) The rings are normalized by program to regular shapes.
(d) The diagram is rotated by 180° about a horizontal axis to give a conventional view and the position of the NH_2 substituent is adjusted.

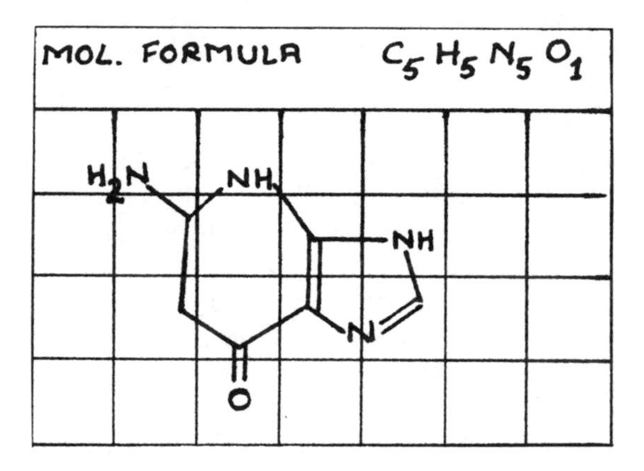

Fig. 4.1 – CSD – Digitizing of hand-drawn copy.

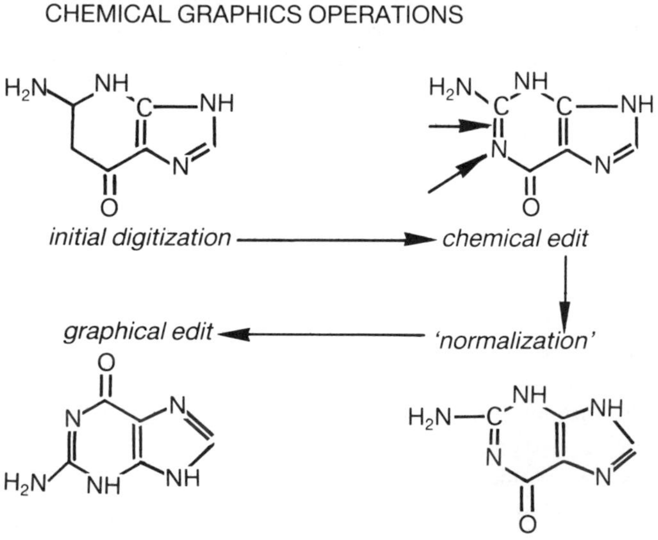

CHEMICAL GRAPHICS OPERATIONS

initial digitization → chemical edit

↓

graphical edit ← 'normalization'

Fig. 4.2 – CSD – Editing of diagram in chemical graphics operation.

Output from the system is recorded as a connection table (Fig. 4.3) in which the x,y-coordinates of atoms in the final diagram are added to the atom properties.

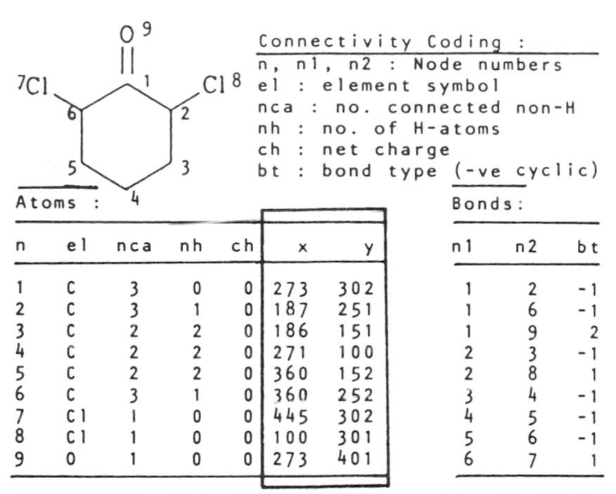

Connectivity Coding :

n, n1, n2 : Node numbers
el : element symbol
nca : no. connected non-H
nh : no. of H-atoms
ch : net charge
bt : bond type (-ve cyclic)

Atoms :

n	el	nca	nh	ch	x	y
1	C	3	0	0	273	302
2	C	3	1	0	187	251
3	C	2	2	0	186	151
4	C	2	2	0	271	100
5	C	2	2	0	360	152
6	C	3	1	0	360	252
7	Cl	1	0	0	445	302
8	Cl	1	0	0	100	301
9	O	1	0	0	273	401

Bonds:

n1	n2	bt
1	2	-1
1	6	-1
1	9	2
2	3	-1
2	8	1
3	4	-1
4	5	-1
5	6	-1
6	7	1

Additional atom properties x, y, are chemical diagram plot coordinates.

Fig. 4.3 – CSD – Connection table.

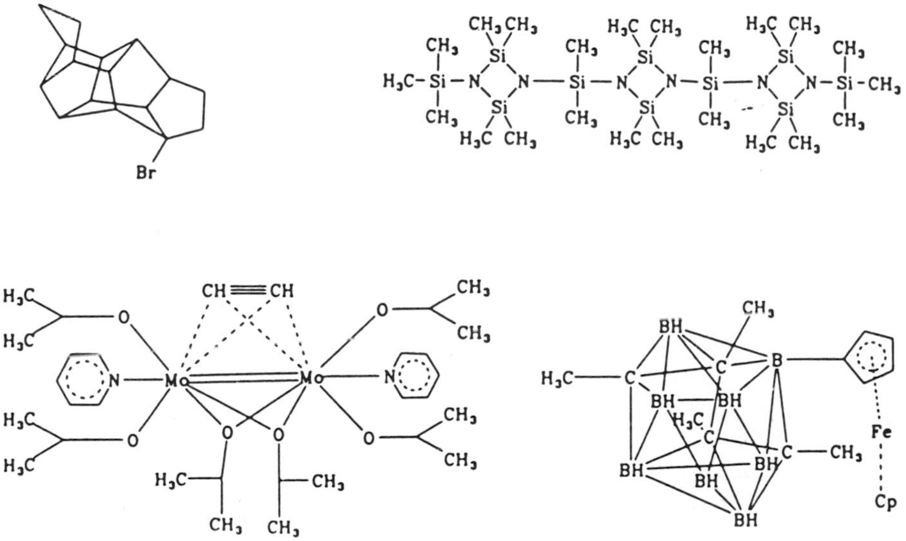

Fig. 4.4 – CSD – Diagram output.

Diagram output has not yet been integrated into CSD software supplied to affiliated centres, but it has been used for production of high quality hard copy for the *Molecular Structures and Dimensions* bibliographies. Thus for Volume 13 (published in 1982) it was possible for the first time to include 3,440 chemical diagrams covering 97.5 per cent of entries cited in the bibliographic listing [7, 8]. Fig. 4.4 shows a selection of more complex diagrams from this compilation to indicate the generality of the system. These include a bridged ring system, an overcrowded molecule, a pi-complex with highly coordinated metal atoms and an example of the treatment of a boron cage.

4.2.3 Quality control

Quality control of data in databanks is required because the data will be used again direct, independently of any circumstantial evidence.

In the Cambridge Structural Database (CSD) of X-ray and neutron diffraction studies of organo-carbon compounds, numeric results are critically evaluated before they are entered in the databank. The primary check involves recalculation of bond lengths from the published cell and atomic coordinates; these are compared automatically with the bond lengths as given by the author in the publication, and stored in CSD specifically for this purpose. Any data entry which fails this and other checks is thoroughly investigated, errors are located and corrected as far as possible, often in conjunction with the author. This internal 'self-checkability' of crystallographic reports is fortunate since more then 15 per cent of papers are found to contain at least one numeric error. In CSD the residual unresolved error rate is less than 1.5 per cent. The CSD is thus more accurate than the primary literature from which it is derived.

The quality of the data in the NIH-EPA Chemical Information System (CIS) is mixed and depends on the source of the file. For example, the CRYST component of CIS is the CSD so the checks on this are as described above.

Mass spectral data are less susceptible to internal checking, but each new spectrum being entered into CIS is examined and, using a derivative of an algorithm proposed by Speck *et al.* [9] a quality index (QI) factor (a number between 0 and 1,000) for the spectrum is calculated. Next, a check of the databank is made to see whether the compound is already represented there. If it is not, the new spectrum and its QI factor are added to the file. If the compound is already represented in the file, only the spectrum with the higher QI value is retained. This procedure does not guarantee that all the entries in the databank are of a particular quality, only that there is one spectrum per compound and that a QI factor is available for each entry.

Thus, whereas the X-ray data are subject to quality control, the mass spectral data are merely 'evaluated'. In a programme at EPA, mass spectra are being measured for commercially important chemicals that are not represented in the databank. Quality control is possible in this case (since spectra are measured) and is being exercised. As a result, all these new spectra are receiving high QI factors as they are added to the databank.

Efforts are in progress to extend the approaches in data evaluation and quality control to the other numeric databanks of CIS. Such progress is relatively straightforward when applied to carbon-13 NMR spectroscopy or infra-red spectroscopy, but becomes more difficult in a field such as toxicology. The RTECS databank is compiled and updated by the National Institute for Occupational Safety and Health (NIOSH). The only control over the quality of the data stems from the fact that NIOSH collects data that have been published in refereed journals. Since the refereeing process is far from certain to detect errors in numeric data, this is not a stringent control; consequently, the toxicology data in RTECS should not be relied on in any quantitative sense. Such data appear almost anecdotal compared to the toxicology data demanded by FDA in connection with any investigational new drug.

The problem of evaluating and controlling the quality of toxicology data is a difficult one. Internal consistency checks generally are not possible and remeasurement of the data is not feasible. Consequently, a 'caveat emptor' approach is taken by NIOSH and, in turn, by CIS. As users enter RTECS, they are greeted with a short general statement concerning the quality of the data. Such statements are provided for every numeric data component of CIS, and, while they do not solve the problem, they do inform the user of possible deficiencies in the data.

4.2.4 Power of databanks

A single item of data or a series of single items of data can be retrieved from traditional reference sources. Computer databanks readily allow the systematic analysis of large numbers of related structures and the data on them. This represents a powerful new research technique capable of yielding results which could not be obtained by any other method. The Cambridge Structural Database is now being used extensively for studies of this nature and progress has been reviewed [10, 11].

This powerful potential of databanks is provided for in the NIH-EPA Chemical Information System where, alongside the files of data, there are programs for three-dimensional conformational analysis and molecular orbital calculations. The MLAB component includes facilities for cluster analysis and pattern recognition studies of numeric data.

It is in these areas that bench chemists (rather than information specialists) are already using online services. The very fact that many databanks offer computational facilities in part explains why databanks are so often used not by intermediaries but by end-users. The end-user has specific subject or technical skills which enable him to do with the data precisely what he wants at the moment of retrieval, a fact which makes the intervention of the information specialist superfluous.

4.3 DEVELOPMENT OF DATABANKS

4.3.1 Funding

There are various methods by which both the development and ongoing costs of producing databanks are met. Some services are provided entirely by the private sector, for example the Dictionary of Organic Compounds (see Chapter 2), and others entirely by government, for example the NIH-EPA Chemical Information System.

Services which are run entirely by government, primarily for government use, are those in which the external user is least well catered for. This is particularly true in the area of discontinuous funding which has affected, for example, the service provided by the NIH-EPA CIS. Money is granted in discrete quanta for both file updates and user support. This means that 'between contracts' there can be irregular file updates and, for example, for 9 months of 1982 there was no user support service.

In the UK several cooperative ventures to produce databanks have been undertaken. These have met with varying degrees of success and the problems encountered are often due to difficulties of funding. In case histories of projects reported by the Royal Society of Chemistry [12] success has only come when government funding has been made available. It is significant that the Mass Spectrometry Data Centre (MSDC) received government funding for 17 years before being expected to become self-financing in 1982. This gives an indication of the time required to develop a successful system, when it is planned as a cooperative venture.

4.3.2 Coordination

The first stage in planning is the identification of a need for that particular file and often this is best done by a group of experts in the field involved. Their expertise is important in knowing what data is available and where, what details should be retrievable and what manipulations are required. The acquisition of the data, both backlogs and new information, is the next consideration. This can cause difficulties when the data required to be brought together is owned by several different organizations. Not only do the rights to the data have to be considered but also the rights to the return on its use. The software required to be used to create the databank and to search it may belong to yet another organization.

The provision of a machine on which to hold the databank and related software can involve a further party requiring a return from the use of the databank. Basic funding is needed for converting data into computer-readable form and for the computer and human resources to create and update the databank. Consideration has to be given to standardization, quality control, validation and correction of the data being put into the databank. Once the databank is available it needs to be marketed and, on a continuing basis, there needs to be user support in the form of special expertise to give training and advice on its use.

A successful databank is often the work of a specialized data centre. The work of three such centres, including two in chemistry – the Crystallographic Data Centre and the Mass Spectrometry Data Centre – has been reviewed recently [13].

4.3.3 Case History: Mass Spectrometry Databank

The development of the Mass Spectrometry Databank can be regarded as a classic story of databank development.

The Royal Society of Chemistry (RSC) took over the Mass Spectrometry Data Centre (MSDC) as a going concern from the Atomic Weapons Research Establishment (AWRE) in 1978. The development of the databank went through the following stages:

(1)	Inception (technical group)	1950s
(2)	Government funding/coordination	1965
(3)	Data collection/*MS Bulletin*	1966
(4)	First edition of *Eight Peak Index*	1970
(5)	Inclusion in MSSS (NIH-EPA CIS)	1973
(6)	Transfer AWRE to RSC	1978
(7)	Termination of grant funding	1982
(8)	Self-supporting	1982–

The impetus for the databank came from a group of mass spectrometrists who in the late 1950s kept in touch with new developments by various informal methods. In 1965 OSTI funded the MSDC, which was set up at AWRE where there were many subject specialists and where most of the initial work on data collection had begun. The data centre collected over 3,000 spectra in 3 years, started production of the *Mass Spectrometry Bulletin* and published the first edition of the *Eight Peak Index* in 1970.

RSC were successful in tendering for the MSDC contract in 1978, since the nature of the work was very close to their existing activities of literature scanning and collection of papers meeting specified criteria. The Mass Spectrometry Data Centre aims to collect together all papers with significant information on mass spectrometry. Selected papers are included in the bibliographic *Mass Spectrometry Bulletin* produced monthly. Papers with significant mass spectral data (i.e. 6 or more peaks and intensities for a particular compound) are tagged and authors of these papers are contacted and asked to submit full spectral data for as many compounds as they have available. Those submitted are integrated into the Mass Spectral Data File. The *Eight Peak Index* is generated from information held in this file and the spectral file is incorporated into the Mass Spectral Search System (MSSS) component of the NIH-EPA Chemical Information System.

4.3.4 Need for government action

In developing a new databank service, the major stumbling blocks are finance and coordination. Because the final outcome is likely to be that much more satisfactory if large-scale *planning and coordination* is done, this would seem to be a case for government action. This has been recognized and documented on at least two occasions: the first SRC innovative symposium [14] in 1976 concluded that there was a need for a mechanism for passing data (and news of the need for data) to and fro between the academic world and industry and 'chided the government none too gently for passing the buck between departments on this issue'.

The discussion forum [15] on promoting more effective use of information, held six years later in Admiralty House, May 1982, concluded in the summary report that there was a need for:

(a) closer liaison between producers of statistical information and government departments, trade associations and the private sector.
(b) government funding for specialist 'index' centres.
(c) identification of a specific part of government machinery as having responsibility for support of new concepts and new proposals in the information field.

Martyn [13] stated recently, in his review of specialized data centres in the UK, 'It would be unwise to develop a national policy for the creation of databanks in areas where there appear to be gaps in provision, but the provision of pump-priming facilities for databank development in cases of demonstrated need could be very rewarding'.

4.4 DESCRIPTIONS OF DATABANKS

The databanks described in this section are necessarily a selection, but the major types are represented. The selection is based largely on those databanks which were described at the conference on which this book is based.

4.4.1 The NIH-EPA Chemical Information System

The NIH-EPA Chemical Information System (CIS) [16–20] is the most comprehensive chemical data system available. A series of databanks is linked by central indexing of the chemicals. Programs for analysis and modelling are available for use with the data. The CIS was developed by cooperating Agencies of the US Government and other organizations and is available from the National Technical Information Service (NTIS), through the Telenet and Tymnet networks.

The databanks that are part of the CIS include files of mass spectra, carbon-13 nuclear magnetic resonance and infra-red spectra, X-ray diffraction data for crystals, and mammalian acute toxicity data. There are bibliographic databases associated with the X-ray crystallography and nuclear magnetic resonance spectroscopy, and these have been included within the CIS. The analytical programs include statistical analysis and mathematical modelling algorithms. For example, the chemical modelling laboratory (CHEMLAB) is a multi-faceted program for modelling chemical structures in three dimensions [21]. Given a two-dimensional structure, it can calculate the energetically most favoured three-dimensional structure. The program has many other estimative capabilities, including calculation of solubilities and partition coefficients, and is used in studies of the biological disposition of chemicals for which few physical properties are known.

The hub of the CIS is the Structure and Nomenclature Search System (SANSS), which allows a search through the linked databanks of structures for occurrences of a specific structure or substructure. Searches for specific structures can be carried out by name, molecular formula and CAS Registry Number, and searches for substructures by name fragment, partial molecular formula and query structure. The following description of a structure search is included here to show the full scope of the CIS system, although a full explanation of substructure searching is not given until Chapter 6. The entry of a query structure in CIS is shown in Fig. 4.5. The fragment and ring probe searches shown in Fig. 4.6 retrieve all compounds in the SANSS file containing the required ring and fragments. A substructure search (see Fig. 4.7) can then be carried out on the file created by intersecting the results of the ring and fragment searches. The example of a retrieved SANSS record in Fig. 4.7 indicates the CIS databanks which contain information on this compound. Much of the data can be displayed directly from the SANSS component using universal commands.

Fig. 4.8 shows the present operational CIS system, which can be viewed, as a network of independent databanks, linked together through the SANSS hub, using the CAS Registry Number as the unique universal chemical identifier for each compound The use of the CAS Registry Number to tag all CIS files was codified in EPA regulation no. 2800.2 in 1975 [22]. With the passing of the Toxic Substances Control Act (TSCA) in 1976, the use of the CAS Registry Number was extended to the TSCA inventory, establishing the link between regulatory data and scientific data both within the CIS and in the literature.

The intercomponent linking of the CIS can be used to help solve specific problems and an example is now described. The system is used to identify a chemical and then to obtain related information about it.

The mass spectrum of an unknown water pollutant has been measured. The base peak (100 per cent intensity) in the spectrum is at m/z 77, and when this is entered into the PEAK search option of MSSS (Fig. 4.9) a total of 7566 spectra in the databank are found to have their base peak at this m/z value. Subsequent m/z values from the spectrum of the unknown are entered until only two references satisfy the search criteria. The pollutant thus is tentatively identified as nitrobenzene, CAS Registry Number 98-95-3. To support this confirmation, the carbon-13 NMR spectrum is retrieved. This is accomplished by the command 'GO CNMR', which transfers the user from MSSS to CNMR and then use of the 'SPEC' option. The CAS Registry Number is used to identify the correct spectrum, which is then printed.

The purpose of this search is to identify substances with the structure that matches the following:

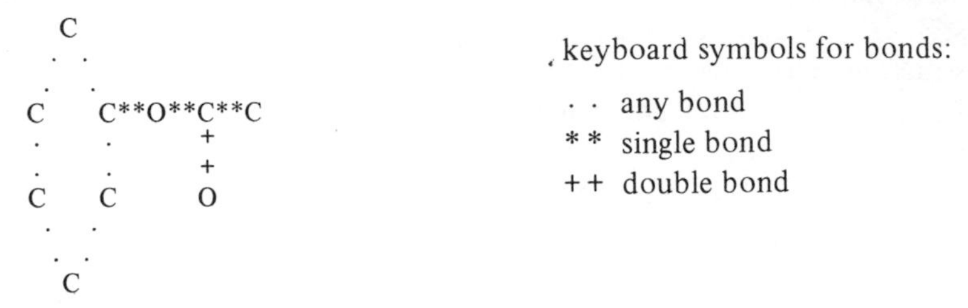

Any halogen may be attached at the para position.

The first step is to construct a query structure on which the search will operate.

Option? **RING**

RING creates a six-membered ring.

Option? **ALTBD 1 2**

ALTBD, for alternating bonds, creates aromatic bonds in the ring.

Option? **ABRAN 1 AT 1**

ABRAN attaches a one-carbon branch at atom 1.

Option? **AGROUP CH3COO AT 4**

```
        10O
        +
        +
        +
8C***9C***11O**4$
```

AGROUP adds an acetate group at carbon 4 of the ring. The computer draws the group added to atom 4. The $ after the 4 indicates that this atom is attached to the rest of the molecule.

Option? **SATOM 7**

Specify up to 10 element symbols = **X**

SATOM will allow multiple atom values to substitute at the atom position specified, in this case atom 7. In place of discreet element symbols, X, the shorthand method of indicating any halogen, has been entered.

Option? **D**

```
      5
     . .
    .   .
   6     4*11O9**8
   .     .     +
   .     .     +
7Z?1     3     10O
     . .
    . .
      2
```

Multi-valued atoms (Z) specified
Node Elements
7 X

Fig. 4.5 – CIS – Entry of a query structure.

RPROBE, or ring probe, is a substructure search program that analyzes a query structure for unique ring systems. The program then selects compounds from the overall SANSS data base that match the query structure in overall structure of the ring system, position and type of heteroatoms, position and type of substituents, and type of bonding at the substituent attachment sites.

Option? **RPROBE**

```
        C??C
      ?      ?
     ?        ?
O**C           C??7
     ?        ?
      ?      ?
        C??C
```

Multi-valued atoms specified
Node Element
7 X

Next, RPROBE prompts for the parameters of the search.

(1) Allow inclusion in larger ring systems (N/Y) (N)? **N**

(2) Allow heteroelements at additional positions (N/Y) (N)? **N**

(4) Allow substituents at additional positions (N/Y) (N)? **N**

Conditions of search

Characteristics to be matched	Type of match
Type of ring or nucleus	EXACT
No heteroatoms	EXACT
Substituents at 4 1	EXACT
Substituents are O *	
Subst bonds are CS A	

This ring/nucleus occurs in 292 compounds.

File = 1, 292 compounds contain this ring/nucleus

Fig. 4.6 – CIS – Search for atom-centred and ring fragments in SANSS – PART 1.

FPROBE, or fragment probe, is a substructure search program that analyzes a query structure for significant atom-centred fragments, and selects compounds from the overall SANSS data base that contain those fragments.

Option? **FPROBE**

FPROBE automatically selects all the atom-centred fragment possibilities from the query structure.

```
Type E to exit from all searches,
T to proceed to next fragment search.

Fragment#  1
          6C.....1C.....2C
1 Multi-valued connected node(s) ignored

Required occurrences for hit  :  6
Fragment#  2
        11O*****4C.....3C
                 .
                 .
                 .
                5C

Required occurrences for hit  :  1

Fragment#  3
        10O+++++9C****11O
                 *
                 *
                 *
                8C

Required occurrences for hit  :  1

Fragment#  4
          9C****11O*****4C
Required occurrences for hit  :  1

Fragment     #2              28812 Hits
Fragments    #2, #4          17622 Hits
Fragments    #2, #4, #3       5246 Hits

File = 2,    5220 compounds contain all 4 fragments
```

Option? #1 **AND #2**

The results of the RPROBE and FPROBE searches are intersected.

File: 3 Count: 77

Fig. 4.6 – CIS – Search for atom-centred and ring fragments in SANSS – PART 2.

The substructure search program, or SUBSS, usually is invoked after FPROBE and RPROBE searches. FPROBE and RPROBE assure only that selected fragments or ring systems appear somewhere in the retrieved compounds. SUBSS will perform a bond by bond, atom by atom match to insure that the query structure is imbedded, exactly as drawn, somewhere in the retrieved compounds.

Option? **SUBSS 3**

A substructure search is performed against file 3 to retrieve those compounds that contain the query structure.

Doing sub-structure search
Type E to Exit

```
File entry —    20, Hits so far    6
File entry      40, Hits so far    9
File entry      60, Hits so far   12
File = 4       Successful sub structures = 15
```

Option? **SSHOW 4**

How many (E to Exit)? **1 S1**

```
Entry        2   CAS RN 122-88-3
CIS Sources Of Information

      2 - CIS, EI Mass Spectrometry
     22 - NBS, Crystal Data File: 122-88-3.01
     32 - NIOSH/CIS, RTECS: AG0175000
     63 - CIS CTCP, Chem Toxicology of Commercial Products: 101
          9
    144 - Code of Federal Regulations(CHEMLAW)

12 Non-CIS References Available
                                                          C8H7CLO3

                 C
                . .
               .   .
              C     C**CL
              .     .
              .     .
O%%C**C**O**C       C
   %         .    .
   %          . .
   O           C

Acetic acid, (4-chlorophenoxy)- (9CI)
Acetic acid, (p-chlorophenoxy)- (8CI)
(p-Chlorophenoxy)acetic acid
(p-Chlorophenoxy)acetic acid (ACN)
(4-Chlorophenoxy)acetic acid
  13 more names available
```

Fig. 4.7 – CIS Substructure search and record retrieved from SANSS.

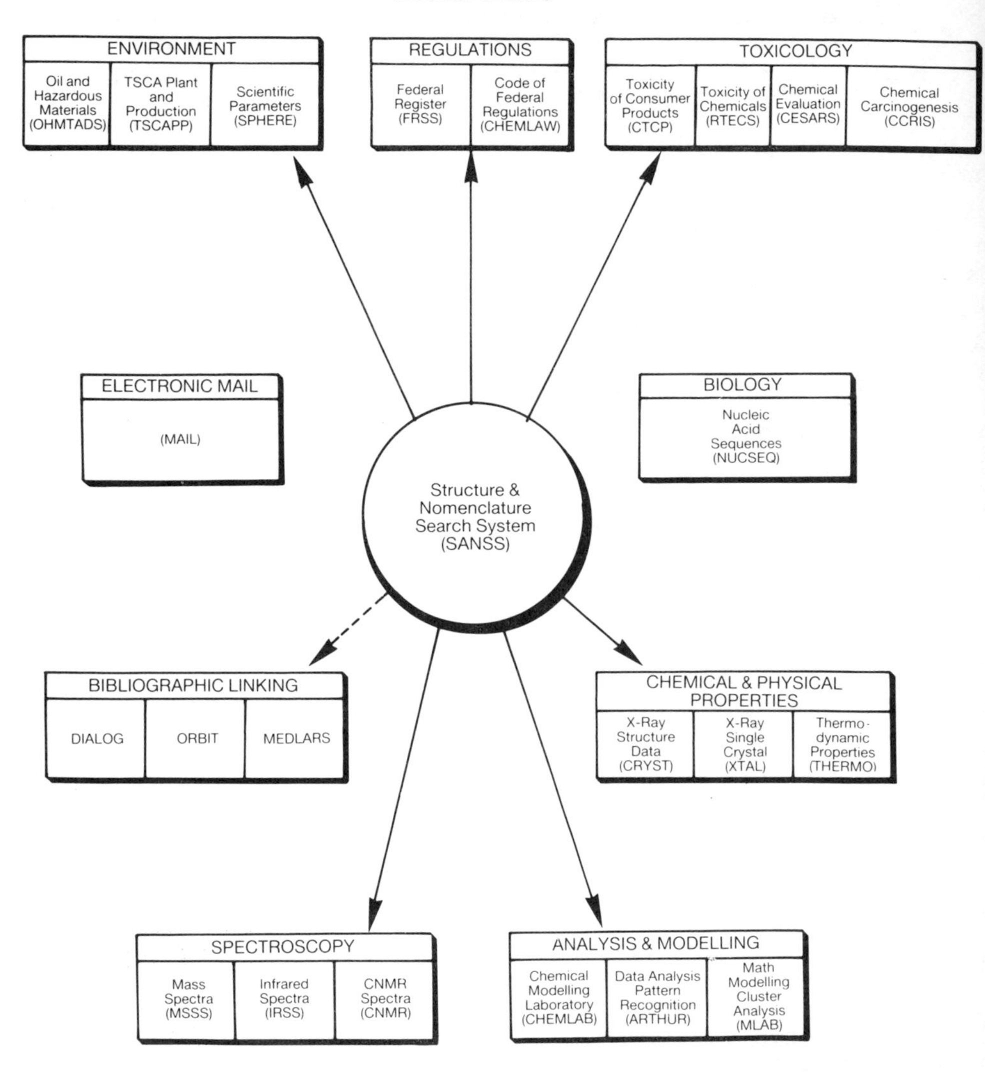

Fig. 4.8 – CIS – Chart of components (full meanings of the acronyms are given in Appendix 1).

```
Option? peak

Type peak,min int,max int
CR to exit, 1 for CAS RN, QI, MW, MF and Name

User:77 10 100

File  1 contains 7566 references to m/z    77

Next request: 51 40 100

File  2 contains  385 references to m/z    77   51

Next request: 123 30 100

File  3 contains   12 references to m/z    77   51  123

Next request: 30 10 100

File  4 contains    2 references to m/z    77   51  123   30

Next request: 1

How many (E to Exit)? 2

Type E to terminate display.
   CAS RN      QI    MW Formula, Names
     98-95-3  716   123 C6H5NO2
                        Benzene, nitro- (8CI9CI)
                        Essence of Mirbane
                        Essence of Myrbane
                        Mirbane (essence of or oil of)
                        Mirbane oil
   5906-99-0  541   217 C6H7N3O4S
                        Benzenesulfonic acid, 2-nitro-, hydrazide (9C
                            I)

Next request:

Option? GO CNMR

Option? SPEC

CAS RN: 98953

          C8
         .  .
        .    .
      C6     C9
       .      .
       .      .
O++N**C4     C7
   +    .    .
   +     . .
   O      C5

Benzene, nitro- (8CI9CI)
WLN: WNR
CAS RN = 98-95-3          CMR#   161       MW = 123.04
C6 H5 N1 O2
C-13NMRSPECTRA,JOHNWILEY&SONS  ,1972,     0
Solvent:    CDCL3

 SHIFT   MULT   INTENS   ASSIGN
 148.2    S        2       4
 134.6    D       49       9
 129.4    D       49       7
 123.4    D       44       5
```

Fig. 4.9 – CIS – Identification of a water pollutant, using MSSS and CNMR.

```
Option? INCAS

CAS numbers (CR to end):  98953

CAS numbers (CR to end):

      1 entries accepted.

File  1 created, contains     1 CAS Registry Numbers.

Option? TSHOW 1

Entry       1    CAS RN 98-95-3          NIOSH number DA6475000

BENZENE, NITRO-
C6-H5-N-O2

SKIN AND EYE IRRITATION DATA:
          skn-rbt 500 mg/24H MOD                  28ZPAK -,61,72
          eye-rbt 500 mg/24H MLD                  28ZPAK -,61,72

MUTAGEN DATA:
          cyt-smc 10 mmol/tube                    HEREAY 33,457,47

TOXICITY DATA:
FO1H30J26 orl-wmn TDLo:200 mg/kg                  ATXKA8 28,208,71
          unk-man LDLo:35 mg/kg                   85DCAI 2,73,70
          orl-rat LD50:640 mg/kg                  AGGHAR 17,217,59
          skn-rat LD50:2100 mg/kg                 GISAAA 24(9),15,59
          ipr-rat LD50:640 mg/kg                  AGGHAR 17,217,59
          scu-rat LDLo:800 mg/kg                  HBAMAK 4,1375,35

DATA REFERENCES:
AGRICULTURAL CHEMICAL
TUMORIGEN
MUTAGEN
SKIN AND EYE IRRITANT
AQUATIC TOXICITY RATING: TLm96:100-10 ppm         WQCHM* 2,-,74
TLV-TWA 1 ppm; STEL 2 ppm (skin)                  DTLVS* 4,303,80
OSHA STANDARD-air:TWA 1 ppm (skin) (SCP-P)        FEREAC 39,23540,74
DOT-POISON B, LABEL:POISON                        FEREAC 41,57018,76

Option? GO OHMTADS

Option? FORMAT CAS,MAT,PRD,HEL

Option? TYPE 1/4


 Conversion to local identifiers resulted in      1 unique occurrences.

        Conversion Entry      1; Accession No.      7216821
 (CAS)  CAS Registry Number: 98-95-3
 (MAT)  Material Name: $$$ NITROBENZENE $$$
 (PRD)  Production Sites: ALLIED CHEMICAL CORP., BUFFALO, NY;
        MOUNDSVILLE, WV;
        RUBICON CHEMICALS INC., GEISMAR, LA;
        E.I. DU PONT DE NEMOURS AND CO., INC., GIBBSTOWN, NJ;
        MONSANTO CO., MONSANTO, IL;
        AMERICAN CYANAMID CO., BOUND BROOK, NJ; WILLOW ISLAND, WV.
 (HEL)  Degree of Hazard to Public Health: HIGHLY TOXIC VIA ALL
        ROUTES. EMITS TOXIC VAPORS WHEN BURNED.
```

Fig. 4.10 – CIS – Retrieval of toxicity data for a water pollutant.

With the identification established, the risk implied by the discovery of this substance in a water supply must be assessed. As shown in Fig. 4.10, the command INCAS permits entry of the Registry Number into the CIS, where it is stored as a temporary file, file 1. Then the command 'TSHOW 1' leads to a transfer from the CNMR area to the RTECS component. Here, lookup, retrieval, and printing of all the toxicity data available for the compound in file 1 take place, and then the user is returned to the CNMR component.

As can be seen, the substance is particularly toxic in most animal species and so, in order to pursue further the risks posed by this substance, a transfer, GO OHMTADS, is requested to the Oil and Hazardous Materials file. The information fields pertaining to production sites (PRD) and degree of hazard to public health (HEL) are sought and the command TYPE 1/4 results in the retrieval and listing of all information in those fields for nitrobenzene.

From the information obtained in this session, it had become clear very quickly that this is a serious pollutant and steps to deal with this situation should be taken immediately.

Much information can be obtained through the chemical locator function of SANSS. The system acts as a directory to over 70 databanks, ten of which are part of CIS. A much more powerful capability of SANSS, however, is its ability to 'package' and 'ship' an entire group of Registry Numbers to other retrieval systems, where the Numbers can be used as search terms. For example, suppose that a substructure search in SANSS results in 400 retrieved compounds. The 400 Registry Numbers can be stored in machine-readable form in a format acceptable as input to DIALOG, SDC's ORBIT, or the National Library of Medicine's database TOXLINE. Once the numbers are stored, the user detaches from CIS, ties in to one of these other systems, and sends the group of Numbers to the new system, where a full search is conducted. In this way, all the literature citations for each of the 400 chemicals can be obtained on demand. This approach combines the structural search capability of CIS with the bibliographic retrieval ability of the other system.

4.4.2 Cambridge Structural Database (CSD)

The Cambridge Structural Database (CSD) was established in 1965 and is a collection of results of X-ray and neutron diffraction studies of organo-carbon compounds published since 1935 [23]. The producer of the database is the Crystallographic Data Centre, Cambridge University, UK. Information in CSD is disseminated by two main routes: via printed books and via machine-readable database releases used in conjunction with suitable software. The bibliographic information is published annually in the reference book series *Molecular Structures and Dimensions* [7, 8]. Cumulative indexes have been published. Numeric data was originally published in book form but this has been discontinued not only because of the massive editorial effort involved, but also since the associated numeric analysis programs are so widely available.

CSD files and programs are implemented at three British centres and at 20 affiliated centres worldwide. Cambridge supplied software operates on the direct sequential files while other systems notably CSSR in the UK [24], the NIH-EPA CIS (see section 4.4.1) in the USA and TOOL-IR [25] in Japan incorporate CSD in other forms, which allow interactive retrieval and display of structures.

The information content of a CSD entry is summarized in Fig. 4.11. Each entry is identified by a reference code, here CLCYHX, and consists of three segments: BIBliography, chemical CONNectivity, and numeric DATA. The chemical connection table for this entry is fully illustrated in Fig. 4.3. The software supplied by Cambridge permits offline searches of BIB or CONN segments followed by retrieval of relevant DATA. This may be processed further to give geometric or graphical display of individual structures, and also provides facilities for the systematic numeric analysis of large numbers of related structures [23].

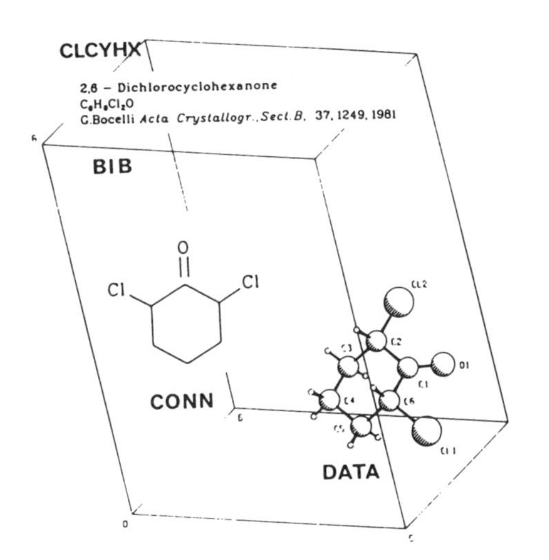

- Organo-carbon compounds only
- No macromolecules
- Complete X-ray or Neutron diffraction study reported

BIB : Literature citation
Compound name(s)
Molecular formula
Chemical classification

CONN : Chemical connection table in compact form

DATA : Atomic coordinates
Cell and symmetry
Accuracy indicators
Text comment

Fig. 4.11 – CSD – Information content of an entry.

4.4.3 European Inventory of Existing Commercial Chemical Substances

In 1967 the Commission of the European Communities issued a Directive (67/548) on the standardization of laws, regulations and administrative provisions relating to the classification, packaging and labelling of dangerous substances.

The 1967 Directive was amended for the sixth time in 1979 by Directive 79/831, which came into effect on the 18 September 1981. This Directive established a framework for the notification of new substances and for the drawing up of an Inventory of substances already on the market.

Under the terms of the 1979 ('6th Amendment') Directive, the physical, chemical, toxicological and ecotoxicological properties of most new substances, either manufactured in or imported into the EEC, are to be adequately investigated so that information on their potential to cause harm to man or the natural environment can be transmitted to an appointed authority within the relevant Member State before the substance is placed on the European market. The only substances completely exempted are those which are already tested under different schemes (drugs, food-stuffs, animal feeds etc.). However, there are various, less stringent requirements and partial exemptions for low tonnage chemicals and to allow the commercial development of chemicals before these are placed on the open European market.

The Directive is concerned with the notification of new substances, not with their approval. Member States are required to monitor 'chemical introductions' and *alert* the appropriate authorities concerning the properties of such substances *before* they come into general use so that any adverse effects are known before the substance is marketed. When a company plans to market a new substance, a dossier of information on the substance is submitted to the Competent Authority in the appropriate Member State. The information is summarized and passed on to the Commission in Brussels and then to the other Member States. The types of information required in the dossier are listed in Table 4.1. Only the main headings are given.

Only new substances which are placed on the market are within the scope of the Directive. 'Existing substances' will eventually be defined as those appearing in the European Inventory, which is an inventory of substances on the market within the European Community between 1 January 1971 and 18 September 1981. Any substances included in the Inventory will, therefore, be outside the notification requirements of the Directive. It has been calculated that it would cost up to £70,000 to provide a comprehensive dossier of information on a new substance. It has therefore been essential to ensure that the Inventory is as complete as possible within the framework of the reporting criteria. These criteria were established by a Commission Decision of 11 May 1981 (81/437/EEC).

Thus, a European Inventory of Existing Commercial Chemical Substances (EINECS) is being prepared and will be published in 1984–85. This will form a definitive list of chemical substances marketed in the EEC and any substance, new or old, which is not included will (subject to specific exclusions) automatically become subject to pre-marketing notification. The Inventory will consist of those substances already listed in the European Core Inventory (ECOIN), discussed below, plus supplementary reports from suppliers or importers of additional substances. This supplementary reporting period ended on 31 December 1982.

The Commission has prepared two basic lists (the Core Inventory and the Compendium) to assist in the task of ensuring that all substances are correctly

Table 4.1 – Notification of new substances: information required for the technical dossier.

1	IDENTITY OF THE SUBSTANCE
1.1	Names (IUPAC) and Trade
1.2	Empirical and structural formula
1.3	Composition of the substances (purity, nature and impurities etc.)
1.4	Methods of detection and determination
2	INFORMATION ON THE SUBSTANCE
2.1	Proposed uses
2.2	Estimated production and/or imports for each of the anticipated uses or fields of application
2.3	Recommended methods and precautions concerning handling, storage, transport, fire etc.
2.4	Emergency measures in the case of accidental spillage
2.5	Emergency measures in the case of injury to persons
3	PHYSICO-CHEMICAL PROPERTIES OF THE SUBSTANCE
3.1	Melting point
3.2	Boiling point
3.3	Relative density
3.4	Vapour pressure
3.5	Surface tension (in aqueous solution)
3.6	Water solubility
3.7	Fat solubility
3.8	Partition coefficient (in *n*-octanol/water)
3.9	Flash point
3.10	Flammability
3.11	Explosive properties
3.12	Auto-flammability
3.13	Oxidizing properties
4	TOXICOLOGICAL STUDIES
4.1	Acute toxicity
4.2	Sub-acute toxicity
4.3	Other effects, including mutagenicity
5	ECOTOXICOLOGICAL STUDIES
5.1	Effects on organisms
5.2	Degradation
6	POSSIBILITY OF RENDERING THE SUBSTANCE HARMLESS
6.1	For industry/skilled trades
6.2	For the public at large

reported. The European Core Inventory (ECOIN) consists of about 33,000 substances arranged alphabetically by CAS preferred name, index name, synonym and, for some substances, trade name. There are separate listings by CAS Registry Number, molecular formula, UVCB subsets and chemical substance definitions. UVCB substances are of Unknown or Variable Composition, Complex Reaction Products and Biological Materials. Examples of such substances are shown in Table 4.2. Chemical substance definitions arranged by CAS Registry Numbers also provide descriptive material on ill-defined substances, as shown in Table 4.3.

Table 4.2 – Examples from the UVCB index.

WOOD PRODUCTS
 Charcoal (16291-96-6*)+
 Cork (61789-98-8*)+
 Pulp, cellulose
  ext., caustic (68424-83-9*)+
  reaction products with calcium silicate (68308-32-7*)
 Pyroligneous acids
  reaction products with Et alc., distillates (8030-89-5*)+
 Waste solids
  pulp mill (68477-27-0*)+

YEAST
 Saccaromyces cerevisias (68876-77-7*)
 Torula
  dried (68602-94-8*)+

ZINC ORES
 Ashes, (residues)
  zinc-refining (69012-86-8*)+
 Calcines
  zinc ore-conc. (69012-79-9*)+

The Compendium contains a further 28,000 substances and is accessed in the same way as ECOIN. Substances listed in the Compendium will also be included in EINECS if they have been reported.

Table 4.3 – Chemical Substance Definitions arranged by CAS Registry Numbers.

9000-24-2*

Galbanum gum
Extractives and their physically modified derivatives
Ferula, Umbelliferae

9000-71-9*

Caseins
A complex combination produced in mammary tissue from amino acids supplied by the blood. It contains several proteins, phosphorus, and calcium.

The procedures described in detail in this section are those for the European Economic Community. Similar procedures were introduced earlier in the USA under the Toxic Substances Control Act [26].

The European inventory described in this section was preceded by a similar inventory in the USA [27]. The file of chemicals in this inventory and the plant and production data submitted are included in the NIH-EPA CIS in the TSCAPP component.

4.4.4 Environment Chemicals Data and Information Network – ECDIN

ECDIN is databank of information on environmental chemicals [28]. An environmental chemical is one which actually or potentially occurs in the environment in significant quantities as a result of human activities and is capable of harming man and other living beings. The databank is produced at the Ispra Establishment of the Commission of the European Communities' Joint Research Centre. ECDIN is available online from the Datacentralen Host Centre in Copenhagen through networks and direct call. The databank contains substance identity information for about 60,000 compounds, acute toxicity data for about 20,000 compounds and more extensive data for 1–2,000 compounds. ECDIN may be likened to an integrated set of computerized handbooks consisting of:

– a chemical dictionary
– a handbook of physical chemical properties
– a directory of chemical producers
– a set of tables containing production and trade statistics
– a handbook of regulations for dangerous substances
– a registry of toxic effects of chemical substances

ECDIN covers a wide range of data related to environmental chemicals and aims to support assessment, prevention and control of the potential hazards linked to the use of a chemical. The databank is organized in files. The names of the files available to the user and some of the data elements contained in these files are shown in Table 4.4.

Table 4.4 – Files and data elements in ECDIN.

Chemical Substances (Registry Numbers)
- ECDIN Registry Name and Number
- Chemical Abstracts Service Registry Number
- The European Customs Union Number
- RTECS Number

Chemical Synonyms (names)

Chemical Structures
- Wiswesser Line-formula Notation
- Chemical Structure Diagram

Chemical Processes

Uses

Production and Trade Statistics

Occupational Health and Safety
- Hazard Information (for man/for working environment)
- General recommendations for human safety
- Personal Protection
- Medical and Biological Surveillance
- Acute poisoning treatment
- Chronic poisoning treatment

Occupational Exposure Limits

EEC Directive 67/548/EEC (dangerous substances – hazard classification)

Classical Toxicity	For each of these, organism, exposure, effects, test condition, comment, bibliographic references
Aquatic Toxicity	
Effects of Microorganisms	

Odour and Taste Threshold Concentrations

Analytical Methods

EINECS The substances included in the European Core Inventory are flagged

Text strings in the data are in the English language but synonyms and a few other data elements are present in six EEC languages (Danish, Dutch, English, French, German and Italian).

Data on chemical economics and on regulations are mainly for the EEC countries, but data concerning the USA and other countries outside the EEC area are included where necessary.

A conversational program has been developed at the Joint Research Centre as part of the ECDIN project specially for the ECDIN databank. The user is guided by a set of menus, which make the dialogue easy to use. The databank can be accessed either through the chemical compound or through the hazard classification from EEC directive 67/548/EEC. In the former case the compound can be searched by one of its names or by one of the numbers assigned to it. When the user has retrieved a compound, or a class of compounds, he selects the data he wants to display from one or more menus. Individual display formats

```
(ECDIN)  27/05/83      Effects On Aquatic Organisms       Menu            Page   1
Substance 0000676  Phenol
                        **** Available Data ****
   ?                                          Count
   A     Medium = Sea Water                      6
   B     Medium = Brackish Water                14
   C     Medium = Growth Medium                  3
Select details required (CR=all) =>
```

If the user replies B the following page will be displayed as the first of 14 possible:

```
(ECDIN)  27/05/83      Effects On Aquatic Organisms      Report 94.01  Page   1
Substance 0000676  Phenol
     Latin Name        : PONTOPOREIA AFFINIS

     Test Parameter    : SWIMMING ACTIVITY
     Effect            : 100.0 % DECREASE
     Comment           : PHOTOPERIOD = 12HR LIGHT AND 12HR IN DARKNESS

     Concentration     : 30.0 MG/L
     Exposure Time     : 0.5-7.0 DAY
                    ---- T E S T   C O N D I T I O N S ----
     Waterflow: Flow Through, Medium: Brackish Water, Place: Laboratory,
     Salinity: 5-6 o/oo,
     Temperature: 4.0 oC, Dissolved Oxygen: >8.0 %,
End Of Page (CR) <BIBLIO> =>
```

Fig. 4.12 – ECDIN – Use of aquatic toxicity file.

have been developed for each file in order to present a complete set of data on a chosen subject as one clear screen picture.

Displaying data on phenol from the Aquatic Toxicity file, the user will see the menu shown in the first part of Fig. 4.12. A reply of 'B' to this selection will result in the display shown in the second part of Fig. 4.12.

4.5 SUMMARY

This chapter has given an indication of the value of the databanks that are playing an increasingly important role in satisfying the chemist's information requirements. Although databanks are concerned mainly with numeric data, their complexity stems from the fact that they also contain text and structural information. Retrieval of information from databanks thus incorporates the bibliographic methods described in Chapters 2 and 3 and also the techniques for structure handling with which the remainder of this book is concerned.

The creation of databanks is very different from the commercial publishing activity of papers in journals, partly because the impetus for compilation of a databank comes from a group of specialists recognizing a need or as a result of legislation. In either case, national or international funding is required at some stage in setting up the databank. The success of a databank thus depends on the enthusiasm of the group responsible for its compilation and on continuing financial support.

REFERENCES

[1] Cronin, B., Databanks, *Aslib Proc.*, **33**, 243–250 (1981).

[2] Rossmassler, S. A., and Watson, D. G., Data Handling for Science and Technology: An Overview and Sourcebook, Amsterdam, North-Holland (1980).

[3] Murdock, J. W., Numerical Data Indexing, *J. Chem. Inf. Comput. Sci.*, **20**, 132–136 (1980).

[4] Walker, S. B., Development of CAOCI and its use in ICI Plant Protection Division, *J. Chem. Inf. Comput. Sci.*, **23**, 3–5 (1983).

[5] de Hamel, C. L., A guide to the published collections of spectral data held by the SRL, 4th edn, Occasional Publications, Science Reference Library, The British Library (1983).

[6] Fisk, C. L., Milne, G. W. A., and Heller, S. R., The Status of Infrared Databases, *J. Chromatog. Sci.*, **17**, 441–444 (1979).

[7] *Molecular Structures and Dimensions*, (Kennard, O., Watson, D. G., Allen, F. H., Bellard, S. A., and Cartwright, B. A. (eds.), Bibliographic Volume 13, Dordrecht, D. Riedel Publishing Company (1982), and other volumes in the series published during 1970–81.

[8] Allen, F. H., Kennard, O., Watson, D. G., and Crennell, K. M., Cambridge Crystallographic Data Centre VI: Preparation and Computer Typesetting of *Molecular Structures and Dimensions* Bibliographic Volumes, *J. Chem. Inf. Comput. Sci.*, **22**, 129–139 (1982).

[9] Speck, D. D., Venhataraghaven, R., and McLafferty, F. W., Computer-aided Interpretation of Mass Spectra. Part XXIII. A quality index for reference mass spectra, *Org. Mass Spectrum*, **13**, 209–213 (1978).

[10] Wilson, S. R., and Huffman, J. C., Cambridge Data File in Organic Chemistry. Applications to Transition-State Structure, Conformational Analysis, and Structure/Activity Studies, *J. Org. Chem.*, **45**, 560–566 (1980).

[11] Allen, F. H., Kennard, O., and Taylor, R., The Systematic analysis of Structural Data as a Research Technique in Organic Chemistry, *Accounts Chem. Res.*, **16**, 146–153 (1983).

[12] Barker, F. H., Development of Chemical Databanks, paper given at the Chemical Structure Association conference 'The Future of Chemical Documentation', Exeter, September 1982.

[13] Martyn, J., Three Specialised Data Centres, *Aslib Proc.*, **35**, 6, 258–277 (1983).

[14] The First SRC Innovative Symposium 1976 'The Industrial Relevance of Recent Developments in the Chemical Thermodynamics of Liquid Mixtures', Leicester Polytechnic, 3–4 November 1976.

[15] Discussion forum on Promoting more Efficient Use of Information, Admiralty House, 13 May 1982 (Report produced by Office of Arts and Libraries with help of Mr. J. E. O. Martyn).

[16] Heller, S. R., Milne, G. W. A., and Feldmann, R. J., A Computer-based Chemical Information System, *Science*, **195**, 253–259 (1977).

[17] Milne, G. W. A., Heller, S. R., Fein, A. E., Frees, E. F., Marquart, R. G., McGill, J. A., Miller, J. A., and Spiers, D. S., The NIH-EPA Structure and Nomenclature System, *J. Chem. Inf. Comput. Sci.*, **18**, 181–185 (1978).

[18] Heller, S. R., and Milne, G. W. A., The NIH-EPA Chemical Information System in Support of Structure Elucidation, *Analytica Chimica Acta*, **122**, 117–138 (1980).

[19] Milne, G. W. A., Fisk, C. L., Heller, S. R., and Potenzone, Jr., R., Environmental uses of the NIH-EPA Chemical Information System, *Science*, **215**, 371–375 (1982).

[20] Heller, S. R., and Milne, G. W. A., On-line Spectroscopic Databases, *American Laboratory*, **12**, 33–34, 38, 40, 42, 44, 46–48 (1980).

[21] Potenzone, Jr., R., Cavicchi, E., Weintraub, H. J. R., and Hopfinger, A. J., Molecular Mechanics and the CAMSEQ Processor, *Comput. Chem.*, **1**, 187–194 (1977).

[22] EPA Order No. 2800.2, issued 27 May, 1975.

[23] Allen, F. H., Bellard, S., Brice, M. D., Cartwright, B. A., Doubleday, A., Higgs, H., Hummelick, T., Hummelink-Peters, B. G., Kennard, O., Motherwell,

W. D. S., Rodgers, J. R., and Watson, D. G., The Cambridge Crystallographic Data Centre: Computer-Based Search Retrieval, Analysis and Display of Information, *Acta Crystallogr.*, B35, 2331–2339 (1979).

[24] Elder, M., Hull, S. E., Machin, P. A., and Mills, O. S., CSSR – Crystal Structure Search Retrieval, User Manual. Daresbury Laboratory, Science and Engineering Research Council, Daresbury, Warrington, UK (1981).

[25] Shimanouchi, T., and Yamamoto, T., Crystallographic Data Services in Japan, Proc. 5th Internat. CODATA Conf., Boulder, Colorado, USA (1976).

[26] Toxic Substances Control Act, USA (Public Law 94-469).

[27] Toxic Substances Control Act Chemical Inventory – Initial Inventory, US Government Printing Office: 1979, 261-699/6179. Environmental Protection Agency, 1979.

[28] Hushon, J. M., Powell, J., and Town, W. G., 'Summary of the History and Status of the System Development for the Environment Chemicals Data and Information Network (ECDIN), *J. Chem. Inf. Comput. Sci.*, **23**, 38–43 (1983).

5

Methods of structure representation and registration

The notion of chemical structure has proved to be of immense importance to the organization of chemical knowledge, and it is on this notion that both manually operated and automated chemical information systems have without exception become established [1–4]. For centuries, chemists have sought means to represent the nature of chemical substances, and the long history of chemical investigation is reflected in part in the multitude and diversity of notational systems [5]. They range from the unsystematic and trivial names for compounds and their reactions, which are still prevalent today because of their ease of written and spoken communication, to the systematic, rule-bound nomenclature schemes used in systematically organized sources of chemical information. The unfolding of the structural theory of chemistry is the result of the pioneering work of many individuals, and it is a tribute to their endeavour that there is today such a powerful and coherent formalism with which to underpin much of modern chemistry.

For the purpose of communicating and recording the identity of a chemical compound, and for providing a key to the indexing of chemical knowledge, the structural formula has proved sufficient. Indeed, the familiar two-dimensional structure diagram has become the common and pervasive vehicle of communication in structural organic chemistry [6]. Modern analytical techniques are such that in many cases the molecular structure of an unidentified compound, for example a synthetic product, can be quickly and indisputably determined, and a unique identification of the compound can be made. Although only a crude approximation to the reality of the chemical molecule, this structural description provides an indispensable key to organizing the vast bulk of chemical knowledge. Chemistry is alone among the scientific disciplines in providing such a key, and it is as a result of this provision that chemistry and its related sciences are comparatively well documented.

This chapter describes the methods of encoding chemical structures for use in computer systems. More detailed accounts of specific instances of these methods are available elsewhere [1–4, 7]. The need for full structure representations to meet the requirements of modern information systems is emphasized, and the different objectives for which these representations are exploited are described. A variety of these objectives may be met within a single system by the interconversion of structure representations into the form most suited for a particular application. The types of interconversion which are possible in principle and in practice are identified. The chapter concludes with a description of compound registry systems, and the role of the registry file within a chemical information system is discussed.

5.1 STRUCTURE REPRESENTATIONS

Several methods for structure representation have been developed. They are discussed with a particular regard to their suitability for computer processing in the context of automated information systems. Each representation derives from the two-dimensional structural formula and is sufficient to provide a simple characterization of that structure. This characterization is ***ambiguous*** if it represents more than a single structure, as in the case of fragmentation codes, or ***unambiguous*** if it represents a single structure only, as in the case of nomenclature, linear notation and connection table representations.

5.1.1 Ambiguous representations

In general, ambiguous representations provide only a partial description of a molecule. The molecular structure is represented generically by a collection of its characteristic fragments, from which reconstruction of the complete molecule is not usually possible. Fragmentation codes, like nomenclature and to a lesser extent linear notations, are classificatory in nature and as such have been used extensively for the indexing and organization of collections of chemical compounds. As a result of its classificatory nature, the effectiveness of a fragmentation code depends critically upon the nature of the compound collection. Accordingly, a fragmentation code is designed so that structural features which are highly characteristic of the compounds in the collection will be prominent, while less common features will be less well represented. If the compound collection is poorly categorized by the code classifications, then the effectiveness of the code in the indexing of and subsequent retrieval from that collection will also be poor. Consequently, a fragmentation code is not generally transportable, without considerable loss of effectiveness, to collections other than that for which is was designed.

Where the nature of the compound collection changes significantly, perhaps as a result of changing interests of the organization in which it is maintained, then a fixed or *closed* code, that is one which consists of a fixed number of fragment terms, may rapidly become obsolete. If the code is assigned intellectually, any modifications made in order to maintain its effectiveness may necessitate the recoding of the entire collection, or the concurrent maintenance of different versions of the code for certain subsets of the collection. In the case of large collections for which only the fragment-coded file exists in machine-readable form, recoding becomes impracticable. Similarly, since the fragment code is also used to search the file, different versions of the code are certain to cause problems for users of the file. Automatic assignment of fragments as search terms is possible where the fragments can be generated from a complete structure representation available in machine-readable form, for example from a connection table. In this case the fragment terms used for searching the file are transparent to the users of the file.

The fragments of an *open-ended* code are constructed from a structure representation by a computer algorithm according to a particular set of rules. Consequently, an open-ended code can be more responsive than a fixed code to the changing characteristics of the compound collection. Also, more precise retrieval from the file may be possible since an open-ended code can offer a much greater variety of search terms than is possible with a fixed code.

The strengths of a fragmentation code for structure representation are its conceptual simplicity and its suitability for use in both manual and automated systems. However, a major problem when searching fragment-coded structure files is a tendency toward imprecise retrieval, or *false drops.* This results from several factors. The most notable of these is the inability to define completely the structural relationships which exist between the assigned fragment terms. Structures may be retrieved which contain the desired fragments but nevertheless do not answer the query, since the fragments do not bear the correct relationship to one another. For these reasons, fragmentation codes are no longer the preferred form of structure representation in automated information systems. Instead, some form of unambiguous representation is invariably used.

Fragmentation codes continue, however, to be used for indexing and searching specialized files, where a code can be specially tailored to achieve maximum effectiveness. Also, fragmentation codes are suitable for indexing structural information from patents in patent documentation systems. Here, structures are expressed generically and are less amenable to other forms of representation [8]. Many systems which are based on other forms of representation generate fragmentation codes automatically for other applications. One such application is structure–property correlation, where there is a need to identify the chemically significant groups in the molecule and to correlate these with known physical and chemical properties. A small number of fragmentation codes have been highly developed for use in certain systems, and these include the SKF (Smith, Kline, and French) code [9], the GREMAS (Generic Retrieval from Magnetic tape Search) code [10] used by IDC (Internationale Dokumentationsgesellschaft für Chemie) [11], the Du Pont code used by IFI/Plenum (Information For Industry) [12], and the Ring Code and the New Chemical Code used by Derwent Publications for their RINGDOC [13] and WPI (World Patent Index) services [14] respectively.

Derwent's New Chemical Code and the GREMAS code are prominent examples of fixed and open-ended codes respectively, and are described in more detail in the following sections.

5.1.1.1 *The Derwent New Chemical Code*

The New Chemical Code was introduced in 1981 as a consolidation of existing codes used within Derwent's patent documentation services. The fragmentation code sections of the New Chemical Code are listed in Table 5.1.

Table 5.1 – Fragmentation code sections of Derwent Publications' New Chemical Code.

Code	Chemical meaning
A to C	Elements present
D to G	Ring systems present
H to L	Functional groups present
M	Overall structure; linking groups etc.
N	Processes; apparatus
P to Q	Activities; properties; uses
R	Formulations: galenical descriptors
S to U	Special steroid codes
V	Natural products; polymers
W	Special dye codes

These sections describe the broad structural classifications and functional group divisions, the description of generic radical substituent groups, and more specific functional group and ring descriptors. The use of the New Chemical Code is illustrated by an online search of the Derwent WPIL (WPI Latest) patents database to retrieve bibliographic references on the bromination of 2-fluoro-4-methyl-biphenyl. Fig. 5.1 depicts the search conducted using the SDC ORBIT retrieval software. Also illustrated is the use of term negation to reduce false drops from the search output.

Negation codes are available for code sections H, J, K and L (Table 5.1) which describe various classes of functional group. Section H deals with common groups which do not contain C=O or C=S, such as nitro, amine, hydroxy, etc. Section J deals with common groups containing C=O and C=S, such as carboxylic acids, esters and amides. Section K covers groups other than NO_2 which contain two heteroatoms linked together, such as N–N, –O–O–, SO_2N, etc. Section L

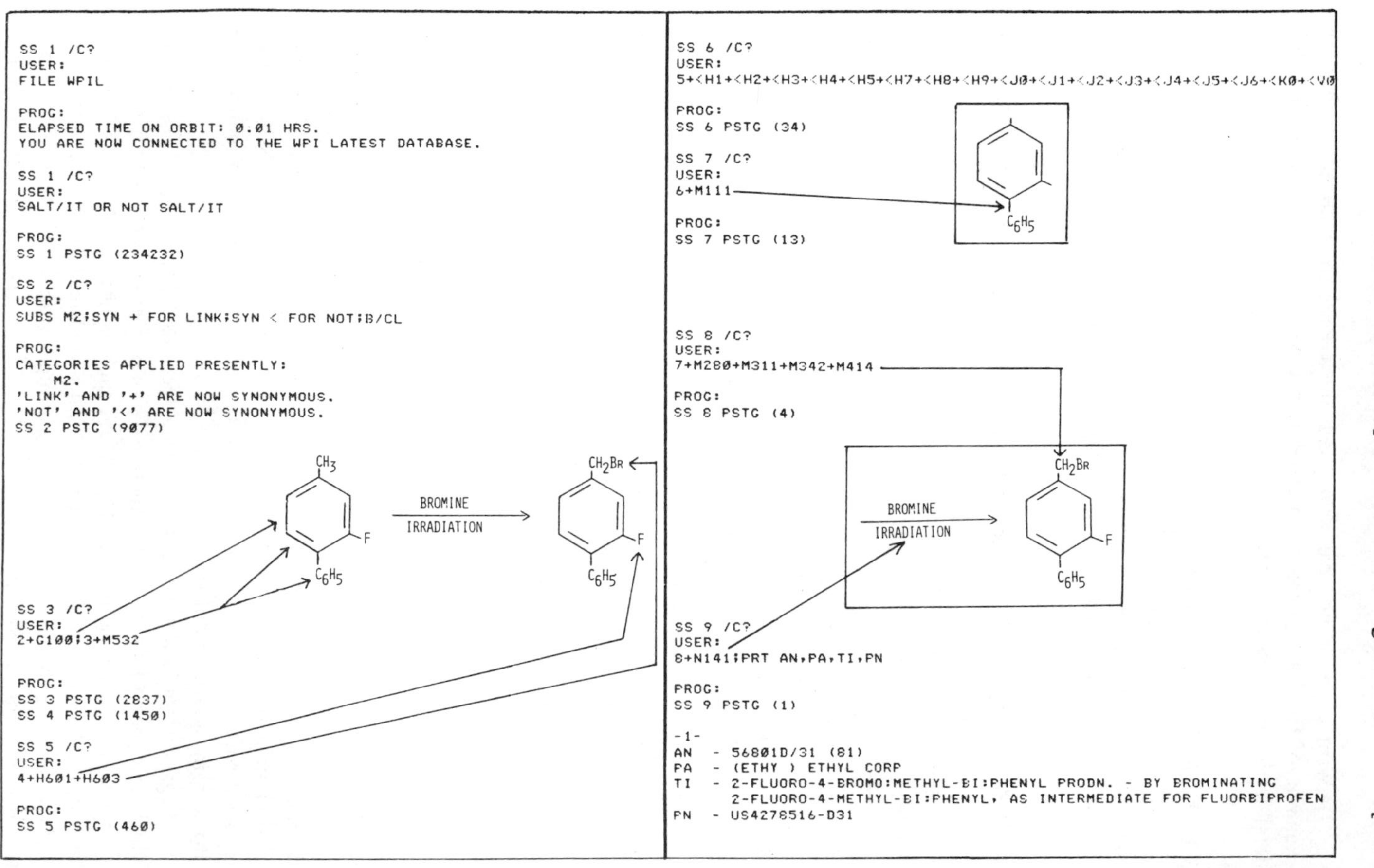

Fig. 5.1 – Searching World Patents Index Latest (WPIL) on SDC ORBIT.

covers other functional groups such as ureas, carbamates and cyanides. Codes for the overall structure of the compound, and linking groups between rings and carbon chains are provided in section M. There are special codes for certain types of compound, for example the steriod (S, T, U), polymer (V) and special dye (W) codes, while other sections provide codes for indexing non-structural information. For example, synthesis and extraction processes are represented in section N.

The first statement SS1 in Fig. 5.1 determines the number of searchable records in the specified file; at the time of search the WPIL file contained 243,232 references to *basic* patents. (The first patent identified in a patent *family* is designated as the *basic*, and all subsequent members of the family are known as *equivalents*. Indexing in WPI for a family of patents is carried out once only on the *basic* patent). SS2 shows that 9077 of these references were in the Farmdoc (M2) section of the WPIL file, which contains pharmaceutical patents. In SS3 and SS4 the codes G100 and M532 specify a benzene ring and two aromatic ring systems respectively and reduce the number of postings from 9077 to 1450. Specification of the codes H601 for fluorine and H603 for bromine in SS5 further reduces this number to 460. SS6 illustrates the use of a series of negation codes combined by the operators + and <, together synonymous with LINK NOT (Boolean AND NOT) logic. These codes further eliminate irrelevant postings and reduce the total to 34. For example, +<H1 eliminates references to structures which contain an amine group, and +<J1 eliminates those containing a carboxylic acid group. Addition of code M111, which specifies a benzene-to-benzene bridging bond further reduces the total to 13, and addition of the codes M280, M311 and M342 for the methylene group and M414 to indicate an aromatic compound reduces the total to 4 postings. Finally, a single reference is obtained by specifying the irradiation process code N141 in SS9.

5.1.1.2 *IDC GREMAS Code*

In the GREMAS code [10], fragment terms are constructed according to a number of broad structural classifications, each of which is subdivided in a hierarchical manner. The classification of each term is described by a three-letter code. For example, the fragment terms for carboxylic acids and their derivatives consist of three-letter codes whose first letter is the *genus* symbol *N*. The second letter of each code qualifies the genus by denoting the *species*, and shows in this case whether these are free acids or specific derivatives, for example esters (*NO*) or amides (*NG*). The GREMAS dictionary contains approximately 700 such genus-species correlations. The third letter of each code qualifies the species by denoting a particular *subspecies*. The subspecies symbol decribes the environment of that species, for example aliphatic amides (*NGA*), and includes designations for various forms of aliphatic, alicyclic, aromatic and heterocyclic environments.

Generalizations within any class of the GREMAS code are described by means of the symbol zero (*0*), and facilitate generic searches. Thus the term *N0A* describes aliphatic carboxylic acids and their derivatives, whereas *NG0* describes carboxamides. Other functional groups and ring systems are described in a similar manner. In the case of ring systems, the first letter of the three-letter code represents the type of ring, specifically its component atoms, the second the ring size, and the third letter the degree of saturation of the ring.

The potential ambiguity of structure encoding using a fragmentation code is illustrated with GREMAS in Fig. 5.2. Here, two distinct structures are assigned identical fragment terms. As noted earlier, this ambiguity will lead to false drops when searching a large structure file, but not to a loss of relevance. With the GREMAS code it is possible to differentiate between the two compounds of Fig. 5.2 by means of a simple grammar. The fragments occurring in each of the partial structures illustrated are listed according to their type and frequency of occurrence. For this purpose, a partial structure is a ring or an uninterrupted chain of carbon atoms. Compound 1 has two partial structures characterized by the term combinations *EN* and *EH*, whereas for compound 2 the corresponding designations are *HN* and *EE*. The means by which terms are combined in this manner constitutes the grammar of GREMAS, for which a syntax is defined. By using this grammar the fragment terms are assigned unequivocally to the corresponding partial structure, with a consequent reduction in the number of false drops. Even if no information is available on the assignment of the fragment terms to the various partial structures, there are means in GREMAS for assigning so-called *syntactic Y-terms*. Use of these terms has been described elsewhere [10].

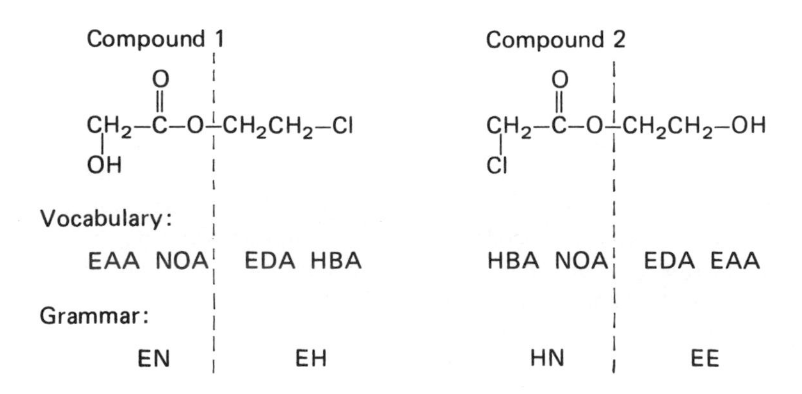

Fig. 5.2 – Vocabulary and grammar of GREMAS.

Recently, algorithms have been developed for the automatic assignment of GREMAS search terms from a connection table. Conversion of Chemical Abstracts Service (CAS) Registry connection tables into GREMAS codes is performed routinely by IDC. It has been found that the specificity of searches based on GREMAS is such that effective searching of files of specific structures is possible using only the fragmented description, without recourse to searching the connection table record.

Fragmentation codes were first designed for manual systems and applied with the primary purpose of organizing collections of chemical compounds. However, the greater emphasis in automated systems is now on the searching of structure files for information retrieval, and for this purpose structures are most suitably recorded in the form of an unambiguous representation. However, the utility of fragmentation codes for indexing generalized structures and classes of compounds is evident from Fig. 5.3, in which successive generalizations of a specific structure are shown. Generalizations of this type occur frequently in chemical patents, in which classes of compounds may be described in the form of a generic *Markush* formula [15]. The scope of the claim is conveniently, if imprecisely, expressed by manual encoding using a fragmentation code, and the major patent documentation services of Derwent, IDC and IFI/Plenum maintain their respective codes for this purpose.

Attempts to extend linear notation and connection table representations in order to encode generic formulae have met with only limited practical success [16]. However, current research in Britain [17] and elsewhere may provide representations which record exactly the scope of a patent claim expressed in this way, and enable precise structure searching of the patent literature for specific compounds and classes of compounds. A brief description of one such approach is given at the end of Chapter 6.

5.1.2 Unambiguous representations

Systematic nomenclature, linear notations and connection tables are all forms of unambiguous structure representation, each capable of representing implicitly or explicitly the connectivity of the complete molecule. The availability of conversion programs has rendered these forms equivalent in many cases, and has resulted in their preferred use for structure representation in today's computer-based chemical information systems. An unambiguous representation may also be a unique, *canonical* representation if the encoding rules or algorithm by which it is created ensure only one preferred encoding for a given structure. In the case of nomenclature and linear notations the canonical encoding, if desired, must be derived intellectually by applying the often complex encoding rules of the nomenclature or notational scheme. It is not generally possible to derive the canonical form algorithmically, nor is it possible to convert a non-canonical form into the canonical form. However, in the case of connection table or similar *topological* representations, automatic interconversion between non-canonical

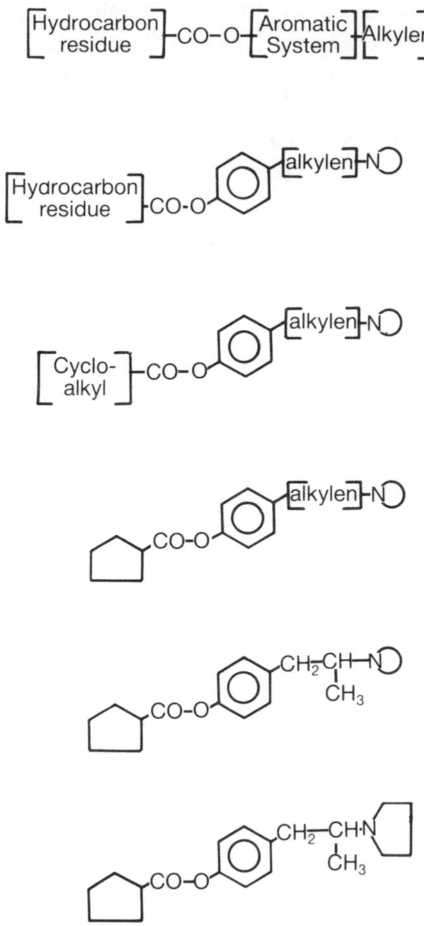

Fig. 5.3 – Generalizations of a specific molecule.

forms and the canonical form is readily possible. The role of the canonical representation in chemical registry systems is discussed in section 5.3.

Of the systematic nomenclatures [18], two are recognized internationally. These are the International Union of Pure and Applied Chemistry (IUPAC) nomenclatures for organic chemistry [19] and inorganic chemistry [20], and the CAS nomenclature [21] applied by CAS nomenclature specialists to ensure a systematic *CA* (*Chemical Abstracts*) Index Name for each new substance appearing in the *Chemical Substance Index* [22] and in *Chemical Abstracts.* An algorithm for automatic generation of *CA* Index Names has recently been reported [23], but not yet implemented in support of the *CA* index preparation. The use of nomenclature as a means of structure representation is extensively documented, and its use in text-based structure searching is discussed in Chapter 6.

Linear notations, like fragmentation codes and many nomenclature systems, predate the general availability of computers. However, certain notations have since been adapted to, and in some cases designed specifically for, computer processing [24]. Only the Wiswesser Line-formula Notation (WLN) [25, 26], for which a user manual is available [27], enjoys an established and widespread use within the chemical community. As a result of its conciseness and relative readability, WLN has become used in handbooks, tables of property values and chemical databanks. Of the major indexing and retrieval tools the *Index Chemicus Registry System* (*ICRS*) [28] and the printed *Chemical Substructure Index* [29] of the Institute for Scientific Information (ISI) are based upon WLN. The CROSSBOW (Computerized Retrieval of Organic Structures Based on Wiswesser) system [30], as its name suggests, supports structure input via WLN and the interrogation of WLN files by string search of the notations.

A connection table is the least structured form of complete representation, and describes explicitly the topology of a chemical structure. In this form, the connection table provides a versatile record capable of supporting the widest range of algorithmic processing typical of large and sophisticated chemical information systems [31]. A connection table is also hospitable to additional structural information, for example stereochemical and tautomeric bonding information, which is less readily incorporated in other forms of representation. Although requiring greater storage than the corresponding linear notation or nomenclature representation, high density storage devices and efficient algorithms for data compaction make it possible to store directly in the form of connection tables large files of chemical structure records. Notable among these is the CAS Chemical Registry System [32], which maintains a connection table and related structural information for each of over 6 million chemical substances indexed in *CA* since 1965.

Direct input of connection tables to the computer is both laborious and error-prone. Chemical typewriters and, more recently, chemical graphics systems have facilitated structure input and enable clerical staff with only minimal training to input two-dimensional structure diagrams for the purpose of database creation. A more complete characterization of a chemical structure is possible where the structure representation makes provision for the description of the stereochemistry of chiral centres, and of atom coordinates which describe the spatial arrangement of the structure. Structure representations in this form are the subject of considerable current interest, notably in three-dimensional substructure retrieval in support of structure–property correlations [33, 34], in the analysis and prediction of chemical synthesis pathways [35], and in the study of molecular interactions in biological systems [36]. These and similar applications are described in Chapter 9.

5.1.2.1 *Linear notations*

The Wiswesser Line-formula Notation (WLN) is representative of a large number of linear notations [24], in which a chemical structure is coded as a sequence of alphanumeric characters. WLN remains predominent among the many notational schemes mainly because of the concerted efforts of its users, continued today through the Chemical Notation Association (CNA), to standardize and regulate the notation [27]. As a string of symbols, WLN lends itself well to use in printed indexes, of which KWIC (Keyword-in-Context) indexes are widely used as bench-top retrieval tools. In addition, WLN is easily adapted to computer processing, both for index production and structure and substructure searching.

The rules which govern the order in which atoms of a structure should be cited in WLN are such that for complex structures, particularly those involving large fused ring systems, encoding into the canonical form can be difficult. Economy is achieved by denoting chemically significant groups and ring systems by a very few symbols. It is estimated that a skilled encoder can encode some 60–100 canonical notations per hour, although a greater speed of encoding can be achieved if the notations are not required to be canonical.

WLN continues to be used in many in-house files for which a complete representation of chemical structure is stored. The reason for this is largely historical. Many of the compound collections began to be indexed in a systematic way at the time of the inception of WLN and similar notations. Indeed, it was the need to organize compound collections in a systematic manner that led to these notations being conceived, and the enthusiasm for the advantages of this form of representation over nomenclature was persuasive. Rapid encoding and the ability to produce sorted and permuted indexes were of considerable advantage. With the advent of computer-based systems, WLN maintained its place by virtue of the economy of structure input and storage, and the suitability of existing text-search algorithms for substructure searching. However, since the introduction of the CAS Chemical Registry System in 1965, there has been a continued change of emphasis towards the connection table as the preferred means of structure representation in both commercial and company information systems. This change has been inspired mainly by the requirements for more flexible substructure search capabilities and the availability of graphics structure input and display systems.

The symbol set of WLN consists of only upper-case alphabetic characters, the digits 0 to 9, and the symbols space, &, / and –. The bonding patterns of atoms and groups are implicit in the notation symbols, and the ordering of these symbols describes unambiguously the connectivity of the structure. Common atoms and functional groups are represented in WLN by a single symbol or a combination of several symbols, the choice of which is further subject to the environment in which that atom or group occurs. For example, the symbol C is reserved for an unbranched carbon atom doubly or triply bonded to at least one other element, as in the nitrile group. Y and X are mnemonic symbols for ternary and quaternary carbon atoms respectively. An unbranched alkyl or alkylene group is represented by a numeral indicating the length of the chain. The symbol O is reserved for a saturated oxygen atom connected to non-hydrogen atoms, as in ethers and esters. As a hydroxyl group, oxygen is represented by the symbol Q, while the carbonyl group, in which oxygen is doubly bonded to carbon, is represented by V.

The choice of symbol depends both upon its structural context and upon the WLN ordering rules, and is illustrated amply in the case of nitrogen atoms. The amino ($-NH_2$) and imino ($-NH-$) groups are represented by the symbols Z and M respectively, while N is reserved for the hydrogen-free atom, as in tertiary amines and the cyano group. In acyclic contexts alone, possible encodings for the urea function (N–CO–N) are ZVZ, ZVM (or MVZ), MVM, NVM, (or MVN), and ZVN (or NVZ), depending upon substitution on either one or both nitrogen atoms. The skilled encoder has little difficulty in choosing the appropriate form. However, this dependency upon context can cause considerable problems for the unwary searcher when framing a substructural query or searching a printed index of notations.

Unsaturations, other than those which are implicit in certain symbols such as V for the carbonyl group, are represented by the symbols U and UU for double and triple bonds respectively, for example

2U3UU1 $CH_3CH{=}CHCH_2C{\equiv}CH$

Branched acyclic structures are encoded by assigning a preferential ordering to each branch and encoding these in turn. WLN symbols are associated with the normal valencies of the common atoms, and as a result the notation fragments which represent distinct branches are uniquely identifiable when terminated by a terminating symbol, for example Z ($-NH_2$). In cases where a terminating symbol is not assigned, the special symbol & is used to represent explicitly the discontinuity in the notation. If the notation continues beyond a terminating symbol or the symbol &, then the origin of the following symbol will be the last symbol of unsatisfied valency.

The description of rings in WLN departs from the philosophy of stating each atom explicitly. Instead, a generalized ring description is given, followed by an indication of the ring size, the position of any non-carbon atoms, the degree of unsaturation, and the means of fusion with other rings. The symbol R is reserved for benzene on account of its frequency of occurrence. Ring systems other than benzene are differentiated between carbocyclic rings and heterocyclic rings. Carbocyclic rings are delimited by the notation symbols L and J, while heterocyclic rings are delimited by the symbols T and J. Letter locants are used to indicate the positions of substituents on rings. Each locant is preceded by a space in order to distinguish it as such, and is followed directly by the symbols for the respective substituent. Several ring notations are illustrated in Fig. 5.4.

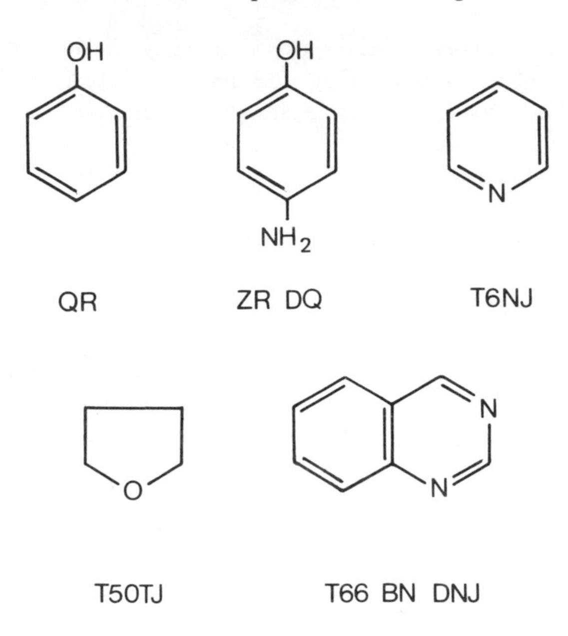

Fig. 5.4 – WLN ring notations.

Although WLN is able to represent unambiguously the large majority of structures, there are instances in which the same canonical notation can be derived for two or more distinct structures, as illustrated in Fig. 5.5. In this case, the notation fragment YS1&1O1 is ambiguous and does not distinguish between the possible connection of the $-CH_2OCH_3$ (1O1) group to either the ternary carbon (Y) or to the sulphur (S) atom. The reason for this ambiguity is that the S symbol represents sulphur atoms of variable valency, and is not included in the set of symbols for which WLN employs the & symbol to make the precise valency explicit. A modified form of WLN, known as ALWIN (Algorithmic Notation Based on Wiswesser), defined on a more rigorous graph-theoretical basis, has been described but has not been adopted in any operational system [37].

Since a canonical WLN provides a unique representation of a chemical structure, WLN has been used for compound registration, as described in section 5.3. In addition, substructure searching is possible using printed indexes and using string search algorithms in automated systems. With the development of the CROSSBOW system at ICI it became possible to translate WLN into a form of *bond-implicit* connection table [38], with which structure display and enhanced substructure searching became possible [30, 39]. This form of connection table is unique to CROSSBOW, and other forms, specifically *bond-explicit* connection tables, have been adopted elsewhere.

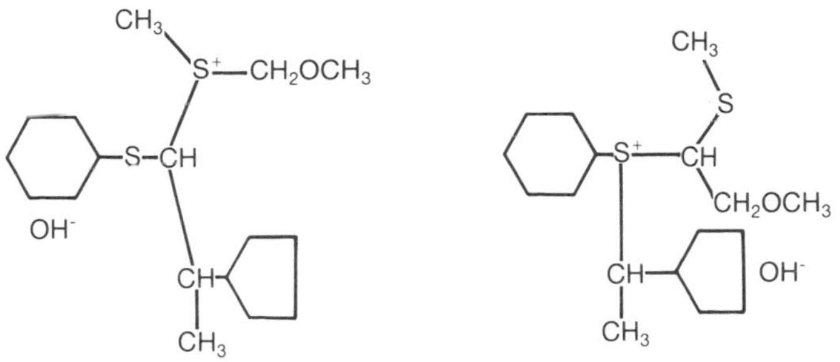

L6TJASYS1 & 101 & Y1 & - AL5TJ & Q

Fig. 5.5 – Ambiguity in WLN.

5.1.2.2 *Connection tables*

Unlike nomenclature and notations, connection tables are seldom input in canonical form, but are converted into such within the computer. In general, the number of valid, unambiguous but non-unique connection tables for any compound equals the number of arbitrary but consistent numberings that can be assigned to the atoms of the structure. A canonical numbering is determined algorithmically, and is used to obtain a unique connection table representation for the purpose of compound registration and storage.

With the exception of CROSSBOW, each of the systems described in Chapter 7 employs a bond-explicit connection table as the principal form of structure representation. In this form, the bonds of the structure are made explicit in the connection table, and the connection table is described as a topological representation. Fig. 5.6 illustrates a connection table for a simple acyclic structure, in which each node is numbered arbitrarily. By convention hydrogen atoms are omitted, and are assumed to fill all unsatisfied valencies. The atom connectivity and bonding information are usually stored in separate tables, but are shown here together for convenience.

A connection table may be stored redundantly, as illustrated in Fig. 5.6, in which case each connection between any two atoms is cited twice, that is once with each of the atoms associated with that bond. More efficient storage can be achieved without loss of information by compacting the connection table so that each connection is cited only once. A non-redundant connection table is illustrated in Fig. 5.7, in which only connections to lower numbered atoms are cited. A file of connection tables may be further compacted by *nesting*, a process by which information common to the representation of two or more related compounds is factored and used as a *heading* for that class of compounds. Nesting achieves a considerable storage economy in large files such as the CAS Chemical Registry file but, in common with any factoring process, recovery of the complete representation can be costly.

$$\begin{array}{c}{}^{3}O \qquad\qquad\qquad\qquad\qquad {}^{8}O \\ \| \qquad\qquad\qquad\qquad\qquad\quad \| \\ {}^{1}NH_2{-}{}^{2}C{-}{}^{4}NH{-}{}^{5}CH_2{-}{}^{6}NH{-}{}^{7}C{-}{}^{9}NH_2\end{array}$$

	Connection	Bond	Connection	Bond	Connection	Bond
1 N	2	1				
2 C	1	1	3	2	4	1
3 O	2	2				
4 N	2	1	5	1		
5 C	4	1	6	1		
6 N	5	1	7	1		
7 C	6	1	8	2	9	1
8 O	7	2				
9 N	7	1				

Fig. 5.6 – Redundant connection table.

$$\begin{array}{c}{}^{3}O \qquad\qquad\qquad\qquad\qquad {}^{8}O \\ \| \qquad\qquad\qquad\qquad\qquad\quad \| \\ {}^{1}NH_2{-}{}^{2}C{-}{}^{4}NH{-}{}^{5}CH_2{-}{}^{6}NH{-}{}^{7}C{-}{}^{9}NH_2\end{array}$$

	Connection	Bond
1N	–	–
2C	1	1
3O	2	2
4N	2	1
5C	4	1
6N	5	1
7C	6	1
8O	7	2
9N	7	1

Fig. 5.7 – Non-redundant connection table.

The choice of connnection table format for a system depends on the use for which the connection table is required. Accordingly, several types of connection table may be required in systems which offer a variety of structure-handling capabilities. For example, a redundant connection table is most appropriate for the purpose of atom-by-atom searching, while a compacted form may be desirable for storage of structure records in large files. Considerations which determine the most appropriate form of connection table for any application have been reviewed previously [31]. Interconversions between these and other forms of structure representation are described in section 5.2.

CAS became the first major user of connection tables when the Chemical Registry System was established. In the current Registry (III) System the unique connection table is still obtained in the form of Fig. 5.7, but is stored in a form in which all acyclic atoms are cited explicitly and each ring is cited only as a Ring Identifier Number (RIN), as illustrated in Fig. 5.8. The connection table for each ring system encountered in the Registry file is maintained in a separate file, linked by the RIN. This reduces the storage for structure records, and facilitates routine operations such as structure drawing [40] and Index Name preparation [41]. Stereochemical information is recorded in the CAS Registry (III) record as a text descriptor. In the MACCS system (Molecular Access System), atom coordinates are stored in the connection table and enable the automatic

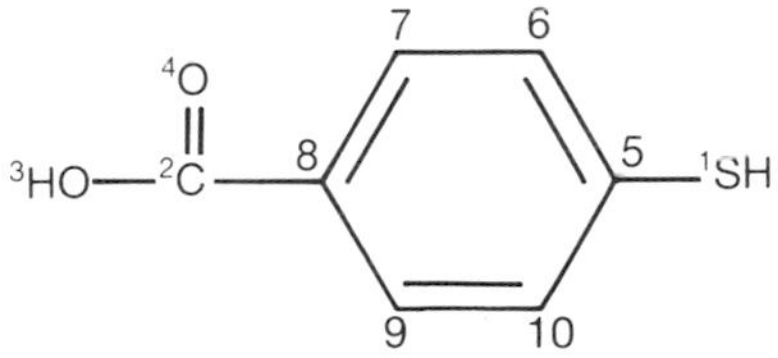

Registry III Connection Table consists of:

(a) Listing of Ring Identifiers

46T.150A.182 (RIN for benzene)

(b) Connection Table for all acyclic atoms and bonds

	Connection	Bond
1S	–	–
2C	–	–
3O	2	1
4O	2	2

(c) Connections between ring(s) and acyclic components

1–5, 2–8

(b) Cross-reference for ring nodes between ring and substance numbering

1	2	3	4	5	6	Ring numbers
5	6	7	8	9	10	Substance numbers

Fig. 5.8 – CAS Registry (III) connection table.

assignment of stereochemical descriptors and the three-dimensional display of structures.

The CROSSBOW connection table combines the chemical significance of WLN with some aspects of a bond-explicit connection table. The chemical significance of the WLN symbols is retained by using a *unit symbol* to represent each atom and its associated bonds. For example, carbon is represented by one of the symbols shown in Table 5.2, the choice of symbol depending on the degrees of substitution and saturation.

Table 5.2 – CROSSBOW unit symbols for carbon.

Symbol	Environment of carbon atom			
L	$-CH_3$, $-CH_2-$			
Y	$-\overset{\vert}{C}H-$			
X	$-\overset{\vert}{\underset{\vert}{C}}-$			
D	$=CH_2$, $=CH-$			
T	$=C<$ (ring)	(endocyclic double bond) ,	$=C<$	(acyclic)
U	$=C<)$	(exocyclic double bond)		
C	$\equiv CH$, $\equiv C-$			
<	$=C=$, $=C:$			
&	alkyl chain of any length			
1–9	alkyl chain of stated length			

The unit symbols are cited in the form of a string. Connectivity between adjacent symbols is assumed unless a discontinuity is specified by means of a *connection transfer*. Fig. 5.9 illustrates the CROSSBOW connection table for an acyclic compound for which the WLN is ZVM1MVZ. The connection transfer 999/009 indicates the end of the connection transfer list (999) and the total number of atoms in the molecule (009). A more complete description of the CROSSBOW connection table is available elsewhere [31], and its use in substructure searching and structure–property correlation has also been described [30].

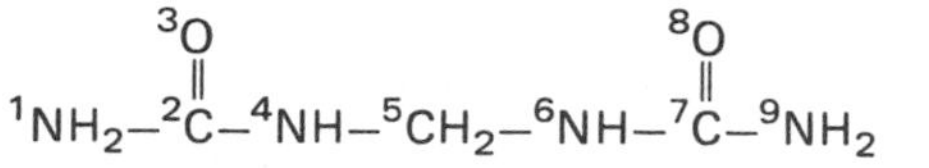

$$\begin{matrix} & & {}^{3}O & & & & {}^{8}O & \\ & & \| & & & & \| & \\ {}^{1}NH_2- & & {}^{2}C- & {}^{4}NH- & {}^{5}CH_2- & {}^{6}NH- & {}^{7}C- & {}^{9}NH_2 \end{matrix}$$

WLN: ZVM1MVZ

Units: MTOMLMTOM

Connection transfers: 003/002 008/007 999/009

Fig. 5.9 – CROSSBOW connection table.

5.2 INTERCONVERSION OF STRUCTURE REPRESENTATIONS

Each of the forms of structure representation described in the preceding section has particular characteristics which recommend its use for certain applications. Classes of compounds, and particularly the *Markush* formulae found in patents, are encoded conveniently by means of a fragmentation code, but this form of ambiguous representation limits the type and effectiveness of structure searching. A linear notation such as WLN provides a concise and unambiguous encoding for many types of chemical structure, but does not of itself favour the flexible search techniques that may be used directly with connection tables in a computer-based system. However, while connection tables may be appropriate for many types of structure manipulations, structure input using WLN is often more economical than graphic structure input.

Interconversion of structure representations may be carried out in order to optimize the effectiveness, and often the efficiency, of a structure-based information system. It is generally possible to interconvert unambiguous representations, in which the connectivity of the molecule is described fully. Similarly, it is possible to convert from an unambiguous to an ambiguous form of representation, such as a fragment screen record for use in substructure searching. Automatic conversion from an ambiguous representation to a more highly organized unambiguous representation, for example from a fragmentation code to a connection table, is not possible. A fragmented record generally contains insufficient information as to the manner in which the various fragments are connected. Interconversion between chemical nomenclature or notations and connection tables, and between different forms of connection table, are the most common types of conversion and are the most important in practice.

Interconversion may be necessary for a number of reasons. It may be required as a result of replacing or upgrading a structure search system for in-house use, or of the need to incorporate a compound file available only in a certain form. For example, installation of a system which uses exclusively connection tables for structure processing will require conversion of existing WLN-based files into the appropriate connection tables. Similarly, conversion of a CAS-derived structure file into CIS (Chemical Information System) connection tables will be necessary before the file can be searched using the NIH/EPA (National Institutes of Health/Environmental Protection Agency) CIS SANSS

(Structure and Nomenclature Search System) search software [42]. In either of these cases, the interconversion is a 'once only' process, and may be achieved in a single batch operation. It may be desirable to take this opportunity to check the accuracy and consistency of structure records either prior to or during the file conversion. This may be done manually by inspection in the case of small files, or may be undertaken by the conversion program.

On a continuing basis, conversion is required where various forms of structure representation are used for different applications within one system. In CROSSBOW, WLN is used for structure input and registration, but is converted into CROSSBOW connection tables for structure searching and display. Structure input to the Cambridge C^{13}NMR database is performed using WLN at the SERC (Science and Engineering Council) Daresbury Laboratory, and is followed by conversion of WLN to NIH/EPA CIS connection tables. Similarly, connection tables for new compounds input to CIS are now being derived by manual encoding from chemical names into WLN followed by conversion into CIS connection tables. This course has been adopted in order to reduce the expense and delays involved with the previous practice of purchasing Chemical Registry connection tables from CAS. A number of notation-to-connection table conversion programs have been developed, but most of these programs are unable to convert a small number of complex ring systems for which the encoding rules are not completely rigorous. WLN-to-connection table conversion programs achieve typically a 90–95 per cent success rate. The remaining compounds must be converted manually. The WLN-conversion program DARING, written at the SERC Daresbury Laboratory to support structure input to CIS, has overcome many of these problems and achieves a negligible failure rate [43].

Automatic conversion from a connection table to a WLN is possible, but published programs have rarely achieved conversion rates of greater than 90 per cent. This is due mainly to the context-sensitive ring descriptions which feature in WLN but which are only implicit in a connection table, and which are therefore difficult to reproduce from the latter. NIH/EPA has published details of a program for converting CIS connection tables to WLN [44], but has not put this into practical use. CAS uses a program developed at Dow Chemicals Co. [45] for generating WLNs of complex polycyclic structures for inclusion in the Parent Compound Handbook [46]. For the same reasons described above, conversion of a connection table into a systematic name is extremely difficult, although the successful implementation of a name generation algorithm which generates *CA* Index Names for a majority of the 1,400 structures indexed per day at CAS, has been reported [41]. The reverse conversion, that is the generation of CAS connection tables from *CA* Index Names [47], is performed routinely by CAS in order to verify the consistency of the connection table and nomenclature records for substances entering the Chemical Registry System.

Conversion of one form of connection table into another is generally possible. The conversion of CROSSBOW connection tables into CAS connection tables is straightforward, while the reverse is complicated due to the explicit ring descriptions in the CROSSBOW connection table. Interconversion of MACCS and CAS connection tables is again simple, but any stereochemical information in the MACCS connection table cannot be represented directly in the corresponding CAS connection table.

5.3 COMPOUND REGISTRATION

An essential function of a chemical information system is the ability to search a file of chemical structures for the presence of a given compound, perhaps represented in the file in a different but equivalent form. This is performed routinely in compound registration, a procedure by which new compounds and accompanying data are added to a structure file and associated data files respectively. A structure file in which a single and unique record of each compound is maintained is known as a registry file. Company registry files are maintained for the purpose of organizing the chemical data generated or gathered within or on behalf of a company, and are used exclusively by that company. Familiar examples of registry files which contain structures reported in the published literature are the *ICRS* [28], and the CAS Chemical Registry System [32].

5.3.1 The Registry Number

It is usual practice to assign a unique identifier to each structure record added to a registry file. In the case of CAS, a Registry Number (RN) is assigned sequentially to each new structure entering the Registry System. In addition to its use throughout the range of CAS publications and services, the RN has also become used in many databanks (CIS, RTECS, ECDIN), primary journals (*Journal of Organic Chemistry*, *Angewandte Chemie*), handbooks (NIOSH, *British Pharmacopoeia*) and compound lists (USAN). The RN is a numeric identifier and has no chemical meaning associated with it. The decision to adopt an identifier of this form was taken at CAS in 1964 during the design of the first Chemical Registry System. The design criteria included the undesirability of imposing upon chemical science a structural classification which, by virtue of the objectives of CAS as a documentary service, would become universal and permanent. As a consequence, a compound and its derivatives, for example its stereoisomers, isotopically labelled isomers, salts and hydrates, are likely to be assigned unrelated RNs. This is especially so if a compound and its derivatives are reported separately over a period of time. For example, the RN of lactic acid is 50-21-5, while the RNs for its S, R and racemic forms are 79-33-4, 10326-41-7 and 598-82-3 respectively. Although these compounds can be retrieved together by substructure search, the case remains strong for a more meaningful registry number which would enable a compound and its derivatives to be brought together in a registry number index.

A variety of alternative registry number formats has been proposed, and these include an identifier which consists of two components. The first of these components would be unique to the topological description of the 'parent' molecule, while the second might designate stereochemical, isotopic and derivative modifications. Searching an index with a registry number truncated after the first component would enable all variations of the basic two-dimensional structure to be retrieved, and a command such as DIALOG's EXPAND, ORBIT's NEIGHBOUR or Data-Star's ROOT would display the registry numbers of all substances with the given 'parent'. As a further stage, the suffix component of the registry number might be designed so as to indicate the exact nature of the modification. Additionally, the first component could be prefixed by a code which identifies the broad category to which the structure belongs. Despite any advantages which might accrue from a structured registry number of this form, it is recognized that the cost of replacing the CAS RN has now become prohibitive owing to its widespread use.

5.3.2 Batch and online registration

The nature of the registration procedure depends upon several factors. These include the following:

(i) the organization of the structure file,
(ii) the nature of the computer system, specifically its batch or online operation, and
(iii) the structure representation employed for input and storage.

Batch-mode registry systems usually employ a serial structure file, against which the batched and sorted file of new structure records is periodically matched. Online registry systems require direct-access to structure files in order to provide for immediate file updating. In both cases, if a structure record is not found at the expected position in the registry file, it is added to the file and assigned a registry number. If additional data is supplied with the substance registered, for example experimental data or a bibliographic citation, then the appropriate data-files are updated. The function of the registry number is to associate these data with that structure record. In an online registry system the structure record and updated data-files are immediately available for searching.

The registry file necessarily contains for each compound a structure representation which is both unique and unambiguous. For this purpose, nomenclature, linear notations and connection tables are suitable. In the case of nomenclature and linear notations, compounds to be registered are encoded into the canonical name or notation, and are matched automatically against the registry file. As a part of the process of input verification, the molecular formula is calculated automatically and is compared with that supplied. The input is rejected if the two molecular formulae are not identical. In the case where the registry file consists of connection table records, structure descriptions may be entered either graphically or as a non-canonical name or notation, from each of which a connection table can be generated.

Numerous algorithms for transforming a connection table into a unique form have been reported. Each of these seeks to assign a unique and invariant ordering to the atoms of the structure, irrespective of the way in which the structure has been drawn or otherwise described. Amongst the simplest of these is that developed at CAS by Morgan [48], which continues to be used in a modified form both in the CAS Chemical Registry System and elsewhere [49, 50]. This class of algorithm partitions the atoms of the structure into classes according to their connectivity. These classes are then refined in an iterative process in which successively higher-order connectivity values are calculated for each atom. Fig. 5.10 illustrates this procedure for a simple molecule.

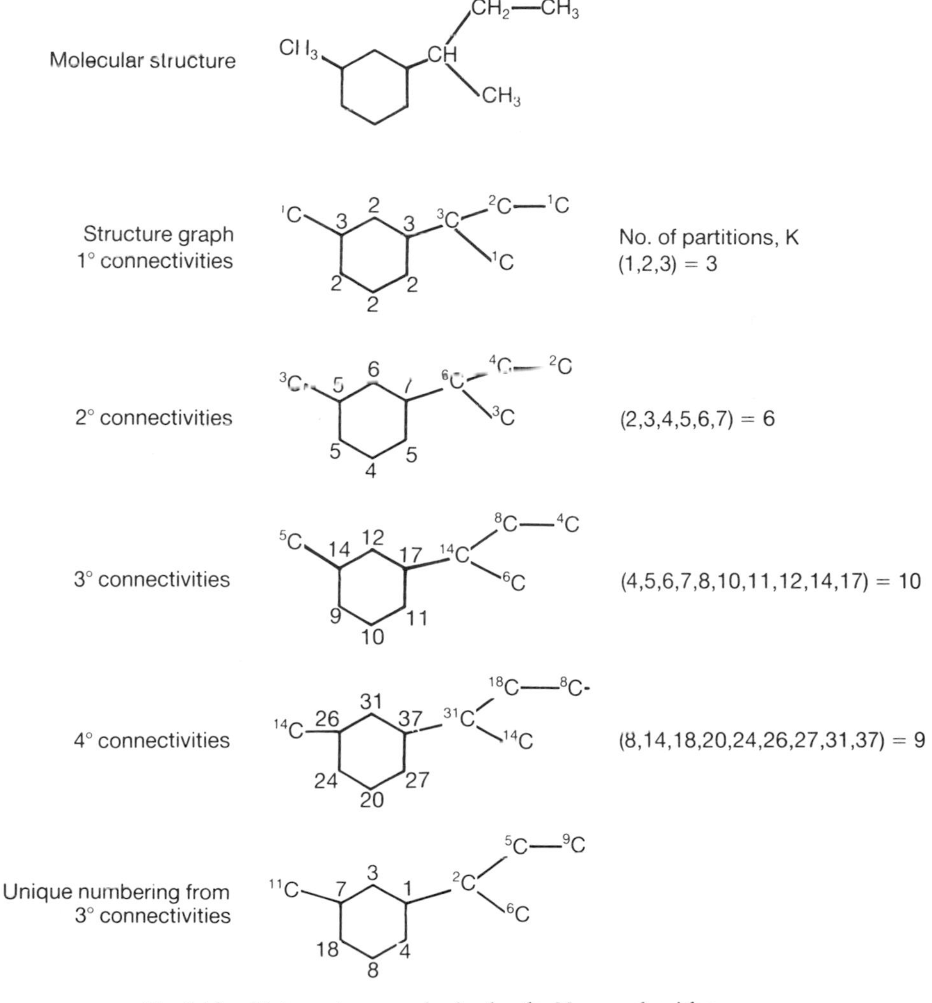

Fig. 5.10 – Unique atom numbering by the Morgan algorithm.

The algorithm terminates after n iterations once the number of classes into which the atoms have been divided ceases to increase. In most cases, each class will contain only one atom and the structure can be numbered uniquely according to the cumulated connectivity values calculated during the n-1th iteration. The atom with the highest cumulated connectivity value is assigned the number 1; its neighbours are numbered 2, 3 etc. in order of their respective values. The unnumbered atoms connected to atom 2 are then numbered, followed by the unnumbered neighbours of atom 3, etc. This procedure is followed until all of the atoms have been numbered. Modifications to the Morgan algorithm take into account additional properties of atoms such as atomic number, the pi-functionality of bonds, double bond isomerism and stereochemistry [49–51].

5.3.3 Efficiency of registration procedures

As shall be seen in the next chapter, direct matching of connection tables, and even of WLNs, is inefficient for large files. A number of methods have been developed for minimizing the time involved in searching a structure file for a given compound. One simple method is the *isomer sort* procedure. Here, the file is partitioned into molecular formula groups with the result that only a small part of the file need be searched for any compound of known molecular formula [52]. However, the distribution of molecular formulae over a file of a heterogeneous nature is highly uneven. As a consequence, some molecular formula groups will be large while others will be small, and the efficiency of the procedure will depend upon this distribution. A variation of the isomer sort technique improves the efficiency of file partitioning by further characterizing the structures in the larger groups by sets of small, bond-centred fragments generated directly from the connection table [53].

An alternative method of file partitioning employs an index code calculated from the structure record, most usually by a *hashing* or key-transformation function [54–56]. This code is matched against the code for each of the structures in the registry file, and an *exact match* search is performed only on those structures with the identical index code. If the structure is already present in the file, it will be found among those structures which have the same index code. In this way, structure comparisons for compound registration can be performed in a time essentially independent of file size. Of course, the same transformation function must be used for registration as was used when creating the file. Computed index codes take a variety of forms, of which three are described below.

5.3.3.1 *ACMF*

The CAS Unique Chemical Registry Record (UCRR) uses four components to represent a compound: a bond-explicit connection table which describes the topology of the structure; a text description concerned principally with stereochemistry; an isotropic labelling component; and a derivative component which, among other functions, describes simple salts and complexes. The Augmented Connectivity Molecular Formula (ACMF) index [54] is computed from the UCRR, and is used both as an index for compound registration, as a direct-access file address and for file validation. The ACMF is very specific, indeed usually unique to a particular compound.

5.3.3.2 *SEMA-name*

A Stereochemically Extended Morgan Algorithm (SEMA) [50], developed for the SECS chemical synthesis program [55, 57], and used in the registration procedures of MACCS, produces a stereochemically unique name from a connection table generated from graphics structure input. The so-called *SEMA-name* is a variable-length string and consists of the following components. A *header* describes the length of the string, the number of non-hydrogen atoms, bonds, double bonds and tetrahedral stereocentres. The *from-list* is an ordered list of all atoms and bonds, except for ring-closure bonds which are listed separately in a *ring-closure list*. A *node list* and a *btype list* describe atoms and bonds respectively in corresponding order. Finally the stereochemical parities for double bonds and tetrahedral stereocentres are specified.

5.3.3.3 *Connectivity index*

In the online registry system developed at Pfizer Research (UK), a connectivity index is calculated from a redundant, non-canonical connection table generated from WLN or graphics structure input [56]. A topological index of this type condenses the connectivity information represented in the connection table into a single numerical index [58]. The index is calculated using a modified Morgan algorithm, which takes into account the different atom and bond types present in the structure.

5.3.4 Compound registry systems

To summarize this chapter, the operation of three online registry systems is described.

5.3.4.1 *CAS Chemical Registry System*

At CAS, each compound encountered during the indexing of chemical publications for inclusion in *CA* is input to the Chemical Registry System [32] by means of any available unambiguous name. The substance name is first matched against the Registry Nomenclature File (RNF) [59]. The RNF is a dictionary file which associates *CA* Index Names with alternative names, molecular formula, ring data, text descriptors and changes in Registry Number. If the name is not found in the RNF, then a structure diagram is prepared and entered by means of an online structure graphics system [60]. Substances for which no connection table is available, for example those of unknown or only partially known structure, and those whose size exceeds the limitations of the Registry System, are filtered out and registered manually [61]. For the remaining compounds, the canonical connection table is generated from the structure diagram using the Morgan algorithm, and is checked for chemical content and edited to produce the UCRR. All cyclic pathways in the structure are determined, and normalized and tautomeric bonds are assigned automatically [62]. The ACMF is computed from the UCRR, and contains stereochemistry and derivative information as appropriate. Structures in the registry file which have the same connection table component of the ACMF are retrieved and the UCRR of each is compared with that of the input structure. If the assigned name but not the UCRR is new, then that name is added to the RNF. If both the name and the connectivity record are new, then a *CA* Index Name is assigned and the appropriate RNF record is created. The UCRR is added to the Registry file and the next available Registry Number is assigned.

The CAS Chemical Registry System now contains over 6.5 million chemical substances reported in *CA* since 1965. CAS has recently begun a programme to expand the Chemical Registry System by including substances cited in *CA* from 1920 to 1965, for which machine-readable structure descriptions were not captured. This programme will result in the further addition to the CAS Registry file of some 1.3 million structure records.

5.3.4.2 *CROSSBOW registration*

The ICI CROSSBOW system is used for the management of chemical databases in a number of organizations. For the purpose of compound registration, a structure is encoded in WLN and entered together with its molecular formula. A check molecular formula is calculated from the WLN and, if this is the same as the molecular formula entered with the WLN, a fragment mask is generated from the notation. The fragment mask is used for a rapid but approximate search of the registry file. Direct comparison of notations is then performed for those structures which match the fragment mask. If the input structure is new, then it is registered by assigning to it a compound registry number, and its fragment mask is stored with that registry number in a separate direct-access search file. If the compond has been registered previously, then the supplied sample, for example, is allocated the next company batch number for that registry number. Data recorded for that sample can then be retrieved using a combination of registry number and company batch number.

5.3.4.3 *Pfizer registration*

In the Pfizer Research (UK) online registry system, structures are entered graphically or in the form of a WLN and molecular formula. In the case of WLN input, validation using a check molecular formula is performed, and the WLN is converted into a connection table. The connectivity index is then calculated from the connection table record, and matched against the registry file and against an extension file containing structure records awaiting addition to the registry file. If a match is found, the corresponding WLNs are retrieved from the registry file and displayed at the terminal for inspection. If no match is found, or if none of the WLNs displayed represents exactly the input structure, then the WLN of the input structure is added to the extension file. The extension file is added to the registry file at appropriate intervals. This system has been designed for an in-house database of some 120,000 compounds, but could be used satisfactorily with considerably larger files.

REFERENCES

[1] Lynch, M. F., Harrison, J. M., Town, W. G., and Ash, J. E., *Computer Handling of Chemical Structure Information*, Macdonald, London (1971).

[2] Wipke, W. T., Heller, S. R., Feldman, R. J., and Hyde, E., (eds.) *Computer Representation and Manipulation of Chemical Information*, Wiley, New New York (1974).

[3] Davis, C. H., and Rush, J. E., *Information Storage and Retrieval in Chemistry*, Greenwood, Westport, Conn. (1974).

[4] Ash, J. E., and Hyde, E., (eds.) *Chemical Information Systems*, Ellis Horwood, Chichester (1975).

[5] Rouvray, D. H., The Changing Role of the Symbol in the Evolution of Chemical Notation, *Endeavour*, **1**, 23–31 (1977).

[6] Russell, C. A., *The History of Valency*, Leicester University Press, Leicester (1971).

[7] Rush, J. E., Status of Notation and Topological Systems and Potential Future Trends, *J. Chem. Inf. Comput. Sci.*, **16**, 202–210 (1976).

[8] Silk, J. A., Present and Future Prospects for Structural Searching of the Journal and Patent Literature, *J. Chem. Inf. Comput. Sci.*, **19**, 195–198 (1979).

[9] Craig, P. N., and Ebert, H. M., Eleven Years of Structure Searching Using the SKF (Smith, Kline and French) Fragment Codes, *J. Chem. Doc.*, **9**, 141–146 (1969).

[10] Rössler, S., and Kolb, A., The GREMAS System, an Integral Part of the IDC System for Chemical Documentation, *J. Chem. Doc.*, **10**, 128–134 (1970).

[11] Fugmann, R., The IDC System, in ref. [4] Ch. 8.

[12] Balent, M. Z., and Emberger, J. M., A Unique Chemical Fragmentation System for Indexing Patent Literature, *J. Chem. Inf. Comput. Sci.*, **15**, 100–104 (1975).

[13] Bawden, D., and Devon, T. K., Ringdoc: The Database of Pharmaceutical Literature, *Database*, **3** (3), 29–39 (1980).

[14] Kaback, S. M., Chemical Structure Searching in Derwent's World Patent Index, *J. Chem. Inf. Comput. Sci.*, **20**, 1–6 (1980).

[15] Valance, E. H., Understanding the Markush Claim in Chemical Patents, *J. Chem. Doc.*, **1**, 87–92 (1961).
[16] Deforeit, H., Caric, A., Combe, H., Leveque, S., Malka, A., and Valls, J., CORA: A Semiautomatic Coding System Application to Coding of Markush Formulas, *J. Chem. Doc.*, **12**, 230–233 (1972).
[17] Lynch, M. F., Barnard, J. M., and Welford, S. M., Computer Storage and Retrieval of Chemical Structures in Patents. Part 1. Introduction and General Strategy, *J. Chem. Inf. Comput. Sci.*, **21**, 148–150 (1981).
[18] Cahn, R. S., and Dermer, O. C., *Introduction to Chemical Nomenclature*, 5th edn., Butterworths, London (1979).
[19] International Union of Pure and Applied Chemistry. Commission on the Nomenclature of Organic Chemistry, *Nomenclature of Organic Chemistry, Sections A, B, C, D, E, F and H*, 1979 edn, Pergamon Press, Oxford (1979).
[20] International Union of Pure and Applied Chemistry. Commission on the Nomenclature of Inorganic Chemistry, *Nomenclature of Inorganic Chemistry*, 2nd edn, Butterworths, London (1971).
[21] Rowlett, R. J., *CA* (*Chemical Abstracts*) Nomenclature, *Chem. Eng. News.*, **53**, (3), 46–47 (1975).
[22] Donaldson, N., Powell, W. H., Rowlett, R. J., White, R. W., and Yorka, K. V., Chemical Abstracts Index Names for Chemical Substances in the Ninth Collective Period (1972–1976), *J. Chem. Doc.*, **14**, 3–14 (1974).
[23] Mockus, J., Isenberg, A. C., and Vander Stouw, G. G., Algorithmic Generation of Chemical Abstracts Index Names. 1. General Design, *J. Chem. Inf. Comput. Sci.*, **21**, 183–195 (1981).
[24] Rush, J. E., Status of Notation and Topological Systems and Potential Future Trends, *J. Chem. Inf. Comput. Sci.*, **16**, 202–210 (1976).
[25] Wiswesser, W. J., *A Line-Formula Chemical Notation*, Thomas Crowell, New York (1954).
[26] Vollmer, J., WLN. An Introduction, *J. Chem. Ed.*, **60**, 192–196 (1983).
[27] Smith, E. G., and Baker, P. A., *The Wiswesser Line-formula Chemical Notation (WLN)*, 3rd edn, Chemical Information Managament Inc., Cherry Hill, New Jersey (1975).
[28] Garfield, E., Revesz, G. S., Granito, C. E., Dorr, H. A., Calderon, M. M., and Warner, A., Index Chemicus Registry System: Pragmatic Approach to Substructure Chemical Retrieval, *J. Chem. Doc.*, **10**, 54–58 (1970).
[29] Granito, C. E., and Rosenberg, M. D., Chemical Substructure Index (CSI). A New Research Tool, *J. Chem. Doc.*, **11**, 251–256 (1971).
[30] Eakin, D., The ICI CROSSBOW System, in ref. [4] Ch. 14.
[31] Ash, J. E., Connection Tables and their Role in a System, in ref. [4] Ch. 11.
[32] Dittmar, P. G., Stobaugh, R. E., and Watson, C. E., The Chemical Abstracts Service Chemical Registry System. I. General Design, *J. Chem. Inf. Comput. Sci.*, **16**, 111–121 (1976).

[33] Gund, P., Three-Dimensional Pharmacophoric Pattern Searching, *Progress in Molecular and Subcellular Biology*, **5**, 117–143 (1977).

[34] Lesk, A. M., Detection of Three Dimensional Patterns of Atoms in Chemical Structures, *Commun. Assoc. Comput. Mach.*, **22**, 219–224 (1979).

[35] Haggin, J., Computers Shift Chemistry to More Mathematical Basis, *Chem. Eng. News*, May 9, 7–20 (1983).

[36] Humblet, C., and Marshall, G. R., Three-dimensional Computer Modelling as an Aid to Drug Design, *Drug Development Research*, **1**, 409–434 (1981).

[37] Krishnan, S., and Krishnamurthy, E. V., Compact Grammar for Algorithmic Wiswesser Notation (ALWIN) using Morgan Name, *Infor. Proc. Man.*, **12**, 19–34 (1976).

[38] Hyde, E., Matthews, F. W., Thomson, L. H., and Wiswesser, W. J., Conversion of Wiswesser Notation to a Connectivity Matrix for Organic Compounds, *J. Chem. Doc.*, **7**, 200–204 (1967).

[39] Thomson, L. H., Hyde, E., and Matthews, F. W., Organic Search and Display Using a Connectivity Matrix Derived from the Wiswesser Notation, *J. Chem. Doc.*, **7**, 204–207 (1967).

[40] Dittmar, P. G., Mockus, J., and Couvreur, K. M., An Algorithmic Computer Graphics Program for Generating Chemical Structure Diagrams, *J. Chem. Inf. Comput. Sci.*, **17**, 186–192 (1977).

[41] Vander Stouw, G. G., Gustafson, C., Rule, J. D., and Watson, C. E., The CAS Chemical Registry System. IV. Use of the Registry System to Support the Preparation of Index Nomenclature, *J. Chem. Inf. Comput. Sci.*, **16**, 213–218 (1976).

[42] Feldman, R. J., Milne, G. W. A., Heller, S. R., Fein, A., Miller, J. A., and Koch, B., An Interactive Substructure Search System, *J. Chem. Inf. Comput. Sci.*, **17**, 157–163 (1977).

[43] Elder, M., The Conversion from Wiswesser Line Notation to CIS Connection Tables, paper presented at CNA(UK) Seminar on Interconversion of Structural Representations, Loughborough University, March (1982).

[44] Farrell, C. D., Chauvenet, A. R., and Koniver, D. A., Computer Generation of Wiswesser Line Notation, *J. Chem. Doc.*, **11**, 52–59 (1971).

[45] Bowman, C. M., Landee, F. A., Lee, N. W., and Reslock, M. H., A Chemically Oriented Information Storage and Retrieval System. II. Computer Generation of the Wiswesser Line Notations of Complex Polycyclic Structures, *J. Chem. Doc.*, **8**, 133–138 (1968).

[46] Ebe, T., and Zamora, A., Wiswesser Line Notation Processing at Chemical Abstracts Service, *J. Chem. Inf. Comput. Sci.*, **16**, 33–35 (1976).

[47] Vander Stouw, G. G., Elliott, P. M., and Isenberg, A. C., Automated Conversion of Chemical Substance Names to Atom-Bond Connection Tables, *J. Chem. Doc.*, **14**, 185–193 (1974).

[48] Morgan, H. L., The Generation of a Unique Machine Description for Chemical Structures: a Technique Developed at Chemical Abstracts Service, *J. Chem. Doc.*, **5**, 107–113 (1965).

[49] Moreau, G., A Topological Code for Molecular Structures. A Modified Morgan Algorithm, *Nouveau Journal de Chimie*, **4**, 17–22 (1980).

[50] Wipke, W. T., and Dyott, T. M., Stereochemically Unique Naming Algorithm, *J. Amer. Chem. Soc.*, **96**, 4825–4834 (1974).

[51] Blackwood, J. E., Elliott, P. S., Stobaugh, R. E., and Watson, C. E., The CAS Chemical Registry System. III. Stereochemistry, *J. Chem. Inf. Comput. Sci.*, **17**, 3–8 (1977).

[52] Lynch, M. F., Orton, J., and Town, W. G., Organization of Large Collections of Chemical Structures for Computer Searching, *J. Chem. Soc. (C)*, 1732–1736 (1969).

[53] Bragg, J. H. R., Lynch, M. F., and Town, W. G., The Use of Molecular Formula Distribution Statistics in the Design of Chemical Structure Registry Systems, *J. Chem. Doc.*, **10**, 125–128 (1970).

[54] Freeland, R. G., Funk, S. A., O'Korn, L. J., and Wilson, G. A., The CAS Chemical Registry System. II. Augmented Connectivity Molecular Formula, *J. Chem. Inf. Comput. Sci.*, **19**, 94–97 (1979).

[55] Wipke, W. T., Krishnan, S. K., and Ouchi, G. I., Hash Functions for Rapid Storage and Retrieval of Chemical Structures, *J. Chem. Inf. Comput. Sci.*, **18**, 32–37 (1978).

[56] Bawden, D., Catlow, J. T., Devon, T. K., Dalton, J. M., Lynch, M. F., and Willett, P., Evaluation and Implementation of Topological Codes for Online Compound Search and Registration, *J. Chem. Inf. Comput. Sci.*, **21**, 83–86 (1981).

[57] Wipke, W. T., Ouchi, G. I., and Krishnan, S., Simulation and Evaluation of Chemical Synthesis – SECS, *Artificial Intelligence*, **11**, 173–193 (1978).

[58] Rouvray, D. H., The Search for Useful Topological Indices in Chemistry, *Amer. Sci.*, **61**, 729–735 (1973).

[59] Callahan, M. V., and Rusch, P. F., Online Implementation of the CA SEARCH File and CAS Registry Nomenclature File, *Online Review*, **5**, 377–393 (1981).

[60] Zamora, A., and Dayton, D. L., The CAS Chemical Registry System. V. Structure Input and Editing, *J. Chem. Inf. Comput. Sci.*, **16**, 219–222 (1976).

[61] Moosemiller, J. P., Ryan, A. W., and Stobaugh, R. E., The CAS Chemical Registry System. VIII. Manual Registration, *J. Chem. Inf. Comput. Sci.*, **20**, 83–88 (1980).

[62] Mockus, J., and Stobaugh, R. E., The CAS Chemical Registry System. VII. Tautomerism and Alternating Bonds, *J. Chem. Inf. Comput. Sci.*, **20**, 18–22 (1980).

6

Substructure search of chemical structure files

Searching for specific chemical substances in computer-based retrieval systems has been addressed in Chapter 5 as part of the procedure for compound registration. For this purpose, determining the novelty of a compound requires the direct comparison of some representation of its structure with that associated with each structure record in the file. This representation may be in the form of a chemical name, for example the systematic *Chemical Abstracts* (*CA*) Index Name assigned according to CAS nomenclature rules, a canonical line notation, or a code computed from the structure record itself, for example by the Morgan algorithm. Among these codes are included the Augmented Connectivity Molecular Formula (ACMF) and the Stereochemically-Extended Morgan Algorithm (SEMA) name, both of which are described in the previous chapter.

Searching for classes of substances related by some common structural feature or combination of features presents a somewhat different problem, and is known as *substructure search* [1]. Substructure search becomes essential where the interest of the searcher lies in the chemistry of the substructure itself; for example, in its involvement in certain types of chemical reaction (see Chapter 8), or in its contribution to the biological activity of a series of compounds (see Chapter 9). In this case, substructure search retrieves from a file every structure which contains the specified combination of substructural features, irrespective of any other features which may be present. The general principles of substructure search are discussed in section 6.1.

Unlike searches for a specific chemical substance, it is not practicable to associate with a substructural query a unique code with which to make direct comparisons with structures in a search file. This is on account of the incomplete specification of attachments to or within the substructure. Accordingly, substructure search has to be carried out by analysis of the stored structure representations, or some parts of these. Since only a small fraction of the structures in a file is expected to satisfy a given substructural query, it is essential to minimize the number of detailed structure comparisons required in the search.

Accordingly, a variety of techniques has been developed, some extensively, with the purpose of achieving at minimum expense the identification of candidate structures from a search file. Factors which influence the design of *screening systems* are discussed in section 6.2.

Substructure search techniques differ according to the chemical structure representation employed in the search file. In the case of nomenclature files, the techniques of text search are appropriate, and chemical names may be searched for desired combinations of name fragments using standard text search packages. Similar methods may also be used to retrieve structures stored as chemical line notations, such as the Wiswesser Line-formula Notation (WLN), for which purpose the notation is treated as text. Substructure search of nomenclature and line notation files is addressed in Sections 6.3 and 6.4 respectively. Connection table representations enable substructure search techniques based on the topology of the structure graph to be employed. Here, the explicit connectivity record allows precise atom-by-atom and bond-by-bond comparisons to be made, in order to confirm the presence or absence of the specified substructure. Substructure search of connection tables is described in section 6.5 in the context of the CAS ONLINE system. This and similar systems search exclusively files of specific chemical substances. The development of new methods for searching files of generic chemical structures, which present many problems of both a conceptual and practical nature, is an important and active area of current research. These problems and the progress towards their solution are summarized in section 6.6.

6.1 PRINCIPLES OF SUBSTRUCTURE SEARCH

Substructure search at the level of detail necessary to confirm the presence or absence of a specified substructure in each structure of a search file can entail a high computational cost. This is especially so in the case of searching connection tables, for which efficient graph matching algorithms have not yet been discovered. Graph and subgraph isomorphism algorithms continue to be the subject of much theoretical study [2], but the nature of these algorithms when applied in the context of searching chemical structures has not changed significantly in recent years. Atom-by-atom search methods have been reviewed previously [3], and particular techniques for substructure search, for example *set reduction*, have been reported in the literature [4, 5].

More recently, a great deal of effort has been put into the design and operation of efficient screening systems. Screening searches are generally classificatory in nature, and seek to associate with each structure a number of substructural fragments, or *screens*, on the basis of which rapid but approximate structure comparisons can be made. Structures which do not share the screens required by the query are quickly eliminated by this search, leaving only those structures which satisfy these requirements to proceed to atom-by-atom search.

It is not easy to distinguish between the function of a fragmentation code (see Chapter 5) when used as a means of substructure search and a set of screens generated by computer algorithms from a structure representation. In each case, structures may be retrieved as candidates on the basis of the presence or absence of the specified structural features, and yet these structures may not be relevant to the query. In the case of a screening search, irrelevant candidates can be eliminated automatically by atom-by-atom search of the stored representation, and complete precision can be achieved, in principle, in every search. In the case of a fragmentation code, the situation is somewhat different. Where the code is assigned intellectually and no complete structure record is retained in machine-readable form, irrelevant candidates can only be eliminated by scrutiny of the search output, since no complete record is available for atom-by-atom search.

Some fragmentation codes, for example the GREMAS code, are designed for either intellectual or automatic assignment and, where automatic assignment from a connection table is made, the terms of the code act as screens in the manner of a screening search. The nature of the fragment terms in a code such as GREMAS differs from those which are designed for exclusively intellectual assignment. In the latter case, for example Derwent's CPI (Central Patents Index) code, the fragment terms represent broad structural classifications and functional group divisions, and will include some number of exact ring and functional group descriptors. Codes designed for algorithmic generation from a machine-readable structure record tend to concentrate on significant atoms, bonds and groups of atoms, and to describe the structural context in which these occur. In the case of the GREMAS code, the computer algorithms which generate these are complex.

Because of practical limitations on their size, screens generated automatically from connection tables are of a more uniform nature than the terms of a manually assigned fragment code. Screens need not correspond to the chemically significant features of a molecule, since they function merely to characterize each structure in order that approximate comparisons can be made as rapidly as possible. In the case of systems which search files of chemical nomenclature and notations, screens will typically include characteristics of the name or notation itself. For nomenclature files these will consist of significant name fragments and codes which describe the microstructure of the name, for example the presence or absence of certain combinations of characters. Where conversion from the name to a connection table is possible, the molecular formula and element counts, and the number, size and identity of rings and other features may be used for screening. For notations, screens will consist of individual symbols and substrings of the notation, as well as other characteristics determined from the notation or, where appropriate, from its equivalent connection table. Examples of screens for substructure search of nomenclature and notation files are given in sections 6.3 and 6.4.

Since a connection table represents a structure explicitly at the atom and bond level, the variety of types of screen which may be generated by computer algorithms is potentially unlimited. However, the types of screen will be determined in practice by the considerations discussed in the following section. Screens generated from connection tables include augmented atoms and a variety of other *atom-centred* and *bond-centred* fragments. These include atom, bond and connectivity sequences of various lengths, and rings detected automatically from the connection table. Screens of these types form the basis of the CAS ONLINE and other substructure search systems and are discussed further in section 6.5. Screens used in the Télésystèmes-DARC system are of different character, and have been described elsewhere [6].

6.2 DESIGN OF SCREENING SYSTEMS

The design of a screening system for substructure search of chemical structure files is influenced by a number of practical and theoretical considerations.

6.2.1 Practical considerations

The practical considerations which influence the design of a screening system and the choice of screens are as follows:

(a) the nature of the structure file
(b) the probable nature of the queries
(c) the characteristics of the use of the file
(d) the size of the file
(e) the file organization.

For structure files of a homogeneous nature, large and specific screens are appropriate where these reflect the major structural characteristics of the file. The likelihood of occurrence of such screens is high, and the combination of a small number of screens can be expected to be highly discriminating. Thus fragmentation codes for collections of structurally related compounds are typified by a small number of large and highly discriminating fragments. For files of a less homogeneous nature which contain structurally diverse compounds, large and specific screens will not correspond well with the types of substructure query which may be anticipated for these files, and which experience suggests will be of a similar diversity [7]. Furthermore, queries are likely to be expressed in more generic terms than substructure queries addressed to a file of closely related compounds. For these reasons, smaller and less specific screens are more suitable for screening heterogeneous structure files. Otherwise, provision must be made for combining many highly specific screens in order to represent adequately the more general features of a substructure query [8]. The generation of screens at various levels of generality is possible only from a connection table, and a number of algorithms have been developed for generating screens at appropriate levels of generality [9]. However, the exhaustive generation of screens by computer algorithm produces ineffective as well as effective screens, with the result that a larger number of these screens needs to be generated in order to provide effective screening. For this reason also, screens of a small size are appropriate if the costs of screen generation are not to bc become impracticable.

The expected use of the structure file for substructure searching will determine the balance between the initial costs expended in establishing and maintaining the screening system and the costs to be incurred in searching the file. Where continuous use of the file for substructure search can be anticipated, this balance will lie in favour of minimizing the recurrent search costs, and particularly of atom-by-atom searching, at the expense of a higher initial investment in the screening system. This investment will take the form of identifying effective screens and maintaining a possibly larger screen set, and in generating and storing these screens for each structure in the file. Where use of the file for substructure search is infrequent, the balance between initial and recurrent costs may lie in favour of minimizing the initial costs of screen generation and storage, at the expense of tolerating longer search times arising from poorer screenout and a greater number of atom-by-atom comparisons.

In the case of a structure file which is relatively small in size, a lower screenout may be acceptable, particularly if an efficient atom-by-atom search is employed, and this can be reflected in the design of the screening system and the nature of the screens. For larger files however, a higher screenout is required in order to maintain as constant the average number of candidates retrieved by the screening search. For example, a screenout of 99.9 per cent will result on average in 100 candidate structures from a file of 100,000 structures. The same screening system applied to the *CA* Registry file of over 6 million structures will result on average in the retrieval of 6,000 candidates for each search. In order to retrieve an average of just 100 candidates from the *CA* Registry file a screening system whose performance is in excess of 99.995 per cent screenout would be required. The expense of maintaining a screening system of such high performance has led to the development of alternative strategies for searching very large structure files, and these are discussed later in this chapter in the context of CAS ONLINE [10].

The organization of the structure file is also an important factor in the design of a screening system. The file organization is often determined by factors other than those which apply to its use for substructure searching. For example, in the case of a compound registry file, the methods of compound registration and of file maintenance and updating, and the interaction with other databases and databanks may decide the choice of file organization. Similarly, an existing database management system may determine the method of file organization. An inverted file organization is used for bibliographic files in most information retrieval systems, and is also commonly used for structure files, particularly fragment-coded, nomenclature and notation files. With certain exceptions, notably the Télésystèmes-DARC and NIH-EPA CIS systems which use forms of inverted file organization (see Chapter 7), a serial file organization is more commonly used for connection table files.

In the case of an inverted file, the structural fragments form an index, in which each fragment is associated with the list of structures indexed by that fragment. The intersection of only those lists which correspond to the fragments of the query enables rapid structure and substructure searches to be performed without the need to search the entire file. The search time depends, among other factors, upon the number of lists which must be intersected in order to achieve a satisfactory screenout, and on the lengths of these lists, that is the number of structures associated with each fragment in the inverted index. Searching may be optimized by intersecting first the shortest lists, since these are associated with the most discriminating fragments, and continuing the intersections selectively until no further significant reductions in the number of candidate structures can be made. Large and discriminating fragments are therefore favoured in an inverted file organization for structure searching. However, these may not always be available for a substructure query, with the result that screenout can be extremely variable depending upon the nature of the query. In addition, search times will vary both according to the number of list intersections and, especially where the screenout is low, to the number of atom-by-atom comparisons which is necessary. Furthermore, file maintenance is costly on account of the very large number of fragments in the index.Updating of the search file is also expensive since the entire inverted index may need to be recreated.

The smaller, generalized fragments which are required to provide a consistently good screening performance over a wide variety of query types are not well suited to an inverted file organization. This is because individual screens will be less discriminating, and the average length of each list in the inverted index will be correspondingly great. Consequently, a serial file organization is appropriate where the screening system is to be based upon small screens. Here the screens assigned to a structure are stored in a bit-mask associated with the connection table. Substructure search involves a serial search of the entire file. Immediate access to the connection table for atom-by-atom search is possible for candidate structures identified by the screens. This enables answers to be displayed as soon as they are found, with an *apparent* search time equal to the time taken to identify the first answer. A serial file organization has been used for medium-sized files and for very large files, for example in COUSIN and CAS ONLINE respectively (see Chapter 7). Screenout and search times depend less critically on the characteristics of the query and, as a result of the novel hardware configurations used in these systems, search times can be held constant as the size of the file increases. Updating of serial files is trivial, and the weekly updates of the *CA* Registry file are immediately searchable through CAS ONLINE.

6.2.2 Theoretical considerations

In addition to these practical considerations, a number of important theoretical considerations influence the design of screening systems for substructure searching. These include the following:

(a) the uneven distribution of substructural features among the structures in a file
(b) the consequence of this distribution on the selection of screens
(c) the interdependence of screens.

Screening searches are achieved most efficiently, in theory, over a broad range of query types if each screen is present in exactly one half of the items in the collection, and if each screen is independent of every other. These principles were first enunciated by Mooers in the context of mechanized information retrieval [11]. It is instructive to note, as Lynch has pointed out [12], that a chemical structure screening system exhibiting these properties would require just twenty screens to distinguish on average one structure from among one million. This ideal cannot be attained in practice for a number of reasons. Firstly, only a small number of characteristics will exist in as many as 50 per cent of the structures in a file, and these are unlikely to be useful as screens. Secondly, the uneven distribution and interdependence of structural features in a structure file is such that a large number of screens needs to be available to provide consistently a satisfactory screening performance. Furthermore, because a substructure query is typically smaller than the structures in the file, the number of screens assigned to the query is generally less than the number assigned to each structure in the file.

The distribution of structural features in a structure file is exemplified by the variation in occurrence of the elements. For example, carbon atoms (72 per cent), together with oxygen (14 per cent) and nitrogen (7 per cent), account for approximately 93 per cent of the total number of non-hydrogen atoms in a typical file of organic molecules. Fluorine, sulphur, chlorine, phosphorus, bromine, silicon and iodine occur less frequently, with iodine typically accounting for less than 0.1 per cent of the total. Distributions of this nature have been recorded for a variety of other structural features [13], and are common elsewhere, for example in the microstructure of text [14]. Thus, in the case of an experimental file derived from the *CA* Registry file, Adamson et al. [15] observed that a small number of atom-centred fragments of a given size occurred very frequently, in somewhat more than half of the file, while relatively few occurred with a moderate incidence. Indeed, the large majority appeared in only few compounds or in just a single compound.

The effectiveness of a set of screens for screening a particular structure file depends in part upon the nature of this distribution. Thus the commonly occurring screens are poorly discriminating, although their utility is increased where they facilitate the encoding of generalized queries. Less commonly occurring screens are more selective but occur only rarely so that their utility for a wide variety of queries is diminished. A balance between these is often sought in practice, and a number of algorithms for generating screens of a much less disparate distribution have been developed [9, 13, 16–19]. These follow one or a combination of two main approaches. The approach developed at Sheffield University in the early 1970s [13–15] seeks an optimal screen set based on the approximate equifrequency of assignment of each screen. This group studied the use of an algorithm which generates a hierarchy of atom-centred and bond-centred fragments, and which increases the size of the commonly occurring fragments in such a way as to produce a screen set in which the incidence of each member in the structure files is appoximately the same. In addition to screens of various sizes, screens of varied atom and bond specificity were also generated. Thus commonly occurring features are appropriately described by screens at a substantial level of detail, while the less common features are represented in more general terms by screens of a lower specificity. Varying the level of description of screens contributes further to the reduction in their uneven distribution, as well as providing general screens to facilitate the encoding of queries for which larger or more specific screens are not appropriate. Bond-centred fragments were found to be preferable to atom-centred fragments for generation by this algorithm since the increase in the variety of bond-centred fragments follows a more even progression through the hierarchy of fragment types. The hierarchies of bond-centred and atom-centred fragments studied by this group are shown in Fig. 6.1.

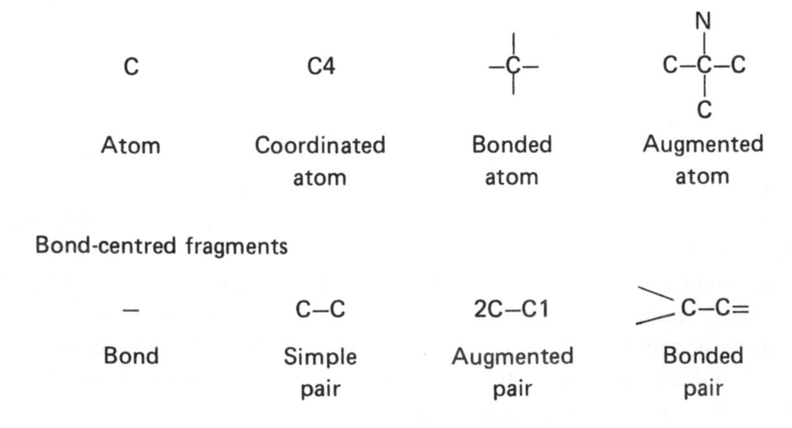

Fig. 6.1 – Hierarchies of atom-centred and bond-centred fragments.

A second approach to screen generation and assignment has been developed for the Walter Reed Army Institute of Research (WRAIR) chemical structure search system [20, 21]. Here, a wider range of substructural sizes is used than considered at Sheffield, but similar procedures are used to increase the size of fragments which occur in the structure file with a frequency above a certain threshold value. Screen generation proceeds by an iterative algorithm, starting with single atoms, in which each iteration adds one atom to each fragment derived from the previous iteration. Fragments are eliminated, or *pruned*, if their frequency of occurrence falls below a threshold of 0.1 per cent and the algorithm proceeds to the next iteration if they occur in greater than 1 per cent of the structures in the file. Fragments of up to eleven atoms have been included in the WRAIR screen set. A number of heuristics are used to prevent the generation of large, highly specific fragments, while ensuring that some larger fragments do remain in the screen set.

The non-independence of screens can lead to a considerable redundancy in a screen set. As a result of this redundancy it is generally not possible to determine the performance of the screening system for a given query on the basis of screen incidence data alone. Redundancy within a screen set has the effect of reducing screenout relative to that which might be expected from incidence data, and arises from a number of circumstances.

Redundancy can arise as a result of the co-occurrence of structural features within a structure. The association between two or more fragments is positive if these fragments occur together more often on average than expected on the basis of a random distribution. The performance of a screen set which includes these fragments will be somewhat less than expected if the screens were independent of one another. For example, Lynch reports a significant positive correlation in one sample file between the elements sulphur and nitrogen, indicating that these elements occur together in rather more structures than would be the case if no correlation existed [12]. Similarly, greater correlations are found to exist among larger fragments. These can be explained, for example in the case of the acyclic simple pairs S=O and C–S, by the fact that these occur frequently together as the fragment C–S=O, which is present in sulphinyl- and sulphonyl-substituted compounds. Negative associations have also been found to exist, although these are less readily interpreted. In contrast to positive associations, the presence of significant negative correlations between screens may improve screening performance for certain types of query.

Also inherent to the structure file itself is the redundancy which arises from the presence within two or more screens of a particularly uncommon structural feature. For example, an element, X, which occurs infrequently in the structure file, will be very selective if used as a screen, but any larger fragment which contains this element will be redundant if included in the screen set. Redundancy arising from this circumstance is often accounted for in the iterative screen generation algorithms which operate on the basis of incidence data. As noted earlier however, these algorithms can introduce further potential sources of redundancy as a result of the lineal relationships which exist between screens. A consequence of these relationships is that if a particular screen is assigned to a structure, every screen derived from it, including those of smaller size and those of a more general nature, will also be assigned [22]. Similarly, if a small screen is assigned, the probability of assignment of one or more larger screens from which the smaller screen is derived is also high, although this probability diminishes as the difference in size between the screens increases.

The pruning mechanism in the WRAIR screen generation algorithm avoids the generation of a great many redundant screens. Nevertheless considerable residual redundancy may remain in a screen set even when generated by this type of algorithm. The WRAIR system incorporates a method of screen assignment by *superimposed* coding [11, 23], which seeks to compensate for this residual redundancy and for the remaining variation in incidence of the screens [20]. Each screen is represented by a pattern of bits in a short bit-mask, rather than by just a single bit in a much longer string. The screen is *weighted* by selecting the number and distribution of bits in the pattern. In theory, the appropriate pattern of bits for each screen can exactly balance the differences in selectivity arising from the variation in incidence among the screens, but only when the screens are independent of one another. A mask of only 96 bits was found to be appropriate for representing in this way each of the 3195 fragment screens in the WRAIR screen set. A screening record for each structure is obtained by superimposing the bit patterns of each of the screens assigned to that structure. The immediate advantage of the superimposed bit string is its evident compactness, which results in reduced storage and a faster screening search. The use of superimposed screens inevitably leads to false drops since some selectivity is sacrificed for compactness. However, theoretical calculations show that the number of false drops which can be attributed to the superimposition of screens is in fact negligible, and in any case these are eliminated by subsequent atom-by-atom search [20].

Comparison of bit strings or of inverted lists for substructure search is made by logical AND and OR operations in either computer hardware or software. In certain circumstances, logical negation (NOT logic) may be used successfully to eliminate structures which contain characteristics expressly forbidden in the description of the query structure. Great care is required when employing a negation search capability at the screening level, especially where the negated feature is represented in the query by the combination of a number of smaller screens. Furthermore, the possibility of negation searching is severely limited in the case of superimposed bit-strings, since the unique correspondence between each bit and each screen is lost [24]. For these reasons, screening searches are commonly based only on the logical AND and OR operations, whilst negation is relegated to the atom-by-atom search stage.

As noted earlier, the screening systems incorporated in many of today's substructure search systems are based in large part on the practical and theoretical considerations outlined in this section. Precise details of screen selection and the balance sought between the cost and performance of a screening system is of interest primarily to the system designers and operators rather than to the searcher, whose interest lies more in the timely and cost-effective retrieval of information.

6.3 SUBSTRUCTURE SEARCH OF CHEMICAL NOMENCLATURE

Manual searching for a specific chemical compound is possible in printed name indexes using the systematic index name and known synonyms, molecular formula and ring indexing information. However, a search in a name index for compounds related by a known substructural feature is considerably more difficult and often leads to less than complete identification of relevant compounds.

The cause of many of the problems associated with substructure search in nomenclature files lies in the scattering of related structures throughout a name index as a consequence of the rules of precedence employed in assigning systematic names. The *CA* Index Names of trimethoprim and a simple analogue are illustrated in Fig. 6.2. The replacement of the 4-methoxy group by a hydroxyl group entails the choice of a different heading parent and the construction of a systematic name that is dramatically different from the first. A search for trimethoprim and its derivatives using significant name fragments is unlikely to be comprehensive unless each of the possible name fragments is enumerated for substances likely to be interest. This is possible for a systematic and well-documented nomenclature system, but is tedious in all but the simplest instances.

Furthermore, retrieval in nomenclature searches is unlikely to be complete due to the variation in the form of name fragments according to their position in a chemical name. Thus, in selecting the search terms to describe a functional group it is necessary to consider not only the case in which that group is the principal function, but also cases in which the group is subordinate to some other functional group. For example, when searching for carboxamides, the search terms 'amino' and 'carbonyl' and the terms 'oxo' and 'aza' should also be considered.

Many text search algorithms permit the association and truncation of name fragments as search terms, and provide for string searching of complete names. These facilities enable substructure searches to be carried out in computer-readable indexes which would be impossible to perform in the equivalent printed indexes [25]. In order to facilitate substance searches in bibliographic files, the major online vendors provide as search aids online chemical dictionaries [26]. Examples of chemical dictionary files are DIALOG's CHEMNAME and CHEMSIS files, SDC's CHEMDEX files, NLM's CHEMLINE file and Télésystèmes' CBNOM.

(A)

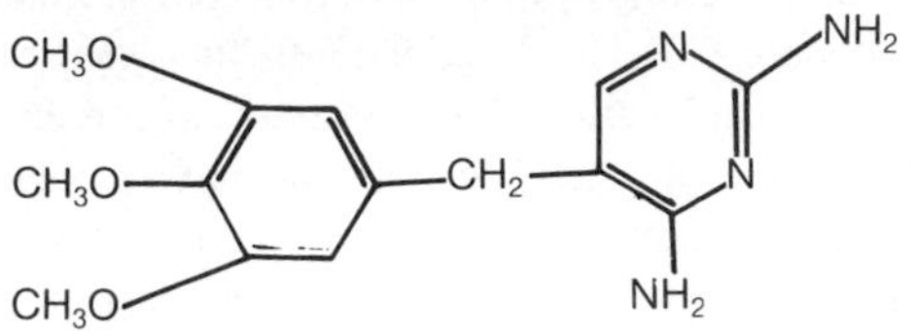

Heading parent: 2,4-pyrimidinediamine

CA Index Name: 5 - [(3,4,5-trimethoxyphenyl) methyl] - 2,4-pyrimidinediamine

(B)

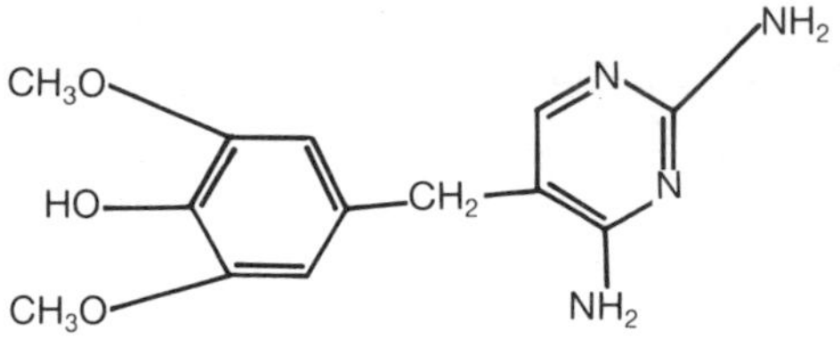

Heading parent: Phenol

CA Index Name: 4 - [(2,4-diamino-5-pyrimidinyl) methyl]) - 2,6-dimethoxy-phenol

Fig. 6.2 – CA Index Names of trimethoprim (A) and a simple analogue (B).

6.3.1 Online chemical dictionaries

Chemical dictionaries perform a variety of functions, the most important of which is the coordination of search terms among the various bibliographic files served by the dictionary. These files may have different naming and indexing conventions and, in addition to systematic chemical names, may use trivial or trade names as indexing terms. A second function is to record the relationships that exist between individual substance records, for example between isomeric forms or among the acid or base salts of a particular compound. The level to which these and other relationships can be described in the dictionary file is limited, with the result that generic searches for compound derivatives, and more generally for compound classes, are more effectively carried out in connection table files.

The large size of some dictionary files can inconvenience those users searching for common or familiar compounds. One approach to solving this problem is to partition the file into a number of smaller files according to the number of times each compound is indexed in the associated bibliographic files. The five

DIALOG dictionary files contain indexing information for over 4.5 million chemical substances indexed by CAS, and are organized by frequency of occurrence in the literature and by *CA* coverage period, as shown in Table 6.1. These dictionary files allow searches to be made on any chemically significant part

Table 6.1 – DIALOG online dictionary files.

File number	File name	CA coverage period	Substances	Occurrences
30	CHEMSEARCH	Vol. 95	Varies	$\geqslant 1$
301	CHEMNAME	Jan. 1967–Sept. 1981	1,201,861	$\geqslant 2$
330	CHEMSIS	Jan. 1977–Sept. 1981	1,381,084	1
329	CHEMSIS	Jan. 1972–Dec. 1976	1,173,442	1
328	CHEMSIS	Jan. 1967–Dec. 1971	842,467	1

of a systematic chemical name. That is, each name is segmented into its basic nomenclatural components by breaking the name at each point of punctuation and subjecting the remaining fragments to an articulation algorithm. Thus the systematic name of trimethoprim, 5-[(3,4,5-trimethoxyphenyl)methyl]-2,4-pyrimidinediamine, is indexed and searchable by the terms, *2*, *3*, *4*, *5*, *trimethoxyphenyl*, *methyl*, *pyrimidinediamine*, *trimethoxy*, *phenyl*, *pyrimidine*, *diamine*, *methoxy* and *amine*. In addition, the DIALOG files also provide access to other substructural features, such as element count, molecular formula, stereochemical descriptors, and analytical information on the number and type of rings present in the compound.

SDC makes available three dictionary files for use with CAS bibliographic files. Details of these files are shown in Table 6.2. The SDC CHEMDEX files, like the DIALOG CHEMNAME and CHEMSIS files, are searchable by chemical name fragments, element count, molecular formula, and stereochemical and ring descriptors. Here, name fragments consist only of those parts of the name

Table 6.2 – SDC ORBIT online dictionary files.

File name	CA coverage period (by RN)	No. of substances
CHEMDEX3	72028-14-9 to 3rd quarter 1980 supplement	281,397
CHEMDEX2	56700-46-0 to 72028-13-8	1,358,125
CHEMDEX	36-88-4 to 56700-45-8	1,638,531

separated by parentheses, commas, colons and hyphens. Accordingly, the *CA* Index Name for trimethoprim is indexed by the terms *2, 3, 4, 5, trimethoxyphenyl, methyl* and *pyrimidinediamine*, but not under the supplementary index terms generated for the DIALOG dictionary files. However, ORBIT string searching enables the CHEMDEX files to be searched for terms embedded within larger index terms, thereby accessing name fragments which are not generated as index terms.

String searching and term proximity operators help to increase the precision of name fragment searching in chemical dictionary and bibliographic files. Name fragment searching lacks precision for essentially the same reasons as those described in Chapter 5 in relation to chemical fragmentation codes, and is primarily a consequence of the lack of context between the fragment search terms. An experimental search system developed in 1971 by CAS and based on the use of *CA* Index Name fragments as search terms for substructure search of the *Chemical Substance Index* identified the major causes of precision failures [27]. Irrelevant substances were retrieved when the required structural features were embedded within the name fragments of larger, unwanted structural units retrieved as a result of search term truncation, for example '*thiazole*' in the terms 'oxathiazole' and 'isothiazole', where '*' represents the truncation operator. Some irrelevant substances contained the required substructural features but in incorrect configurations and carrying unwanted substitutions. A further significant problem identified in this study was the difficulty of searching for substructures which are smaller in size or not related directly to the structural units which make up the vocabulary of terms from which *CA* Index Names are constructed. In this case, nomenclature searching requires the use of a large number of search terms in order to represent adequately each of the substructures of interest.

A limited, batch-mode nomenclature search service was launched by CAS in 1976 after refinement of the earlier experimental system by the addition of more sophisticated text search capabilities [28]. A screening search was also included to improve search efficiency. The screen record for each substance comprises three components: a molecular formula screen, a ring screen, and a nomenclature screen derived from the *CA* Index Name. For those candidate structures identified by the screen search, text search procedures are applied to the full Index Name and, if desired, to related substance data such as molecular formula and ring descriptors. The molecular formula provides an effective screen, particularly if the query substructure contains relatively infrequently occurring elements. For example, sulphur atoms occur in only 20 per cent of substances indexed in *CA*, with the result that searching the molecular formula for the presence or absence of sulphur, and any other infrequent atoms, rapidly eliminates irrelevant substances. The ring screen indicates the presence of certain ring features identified from the substance name, and includes the description of component rings and the number and sizes of these rings. The nomenclature screen is derived from the character pairs which occur in the *CA* Index Name, and each character pair is represented in a short superimposed bit-string [29].

Substructure search in bibliographic files has the particular advantage that substance and non-substance search requirements may be conbined directly within a single query formulation. For example, structural and concept terms may be combined in the form DIBENZO(B,E)(1,4)DIOXIN AND CHLORO AND HAZARD/SAFETY, and executed using text search procedures in each of the appropriate dictionary and bibliographic files. An aspect which is common to all cross-file searches between either chemical structure or dictionary files and bibliographic files is the need to transfer to the latter the registry numbers of the substances retrieved from the former. In most cases this transfer is fully automatic, but nevertheless can be cumbersome if many of the registry numbers are derived from the same document, which is not uncommon in the case of patents and in reports of pharmaceutical research. The DIALOG feature MAPRN allows the searcher to select from among the registry numbers those of particular interest for searching in the bibliographic files, and automatically generates and executes the appropriate search strategy to retrieve bibliographic information on the selected compounds [30]. The SDC ORBIT option PRINTSELECT performs a similar function.

The search techniques used for substructure search of notations are similar to those used for nomenclature files, and are described in the following section. These techniques differ significantly from those employed for substructure search of connection tables, but common to each is the requirement for effective screening, for which the objective remains the same in all cases. A comparison of correlative nomenclature-text and structure-text searching, which discusses factors pertinent to both approaches to substructure search, has been published [31].

6.4 SUBSTRUCTURE SEARCH OF CHEMICAL NOTATIONS

Until recently, many organizations maintained their chemical compound files using WLN as the principal form of complete structure representation. WLN affords a compact storage record, and files of notations can be processed and searched as text files. By rotating notations around their chemically significant symbols and sorting the resulting symbol strings, related compounds can be grouped together within substructure indexes. These indexes can be used for both manual and computer-based substructure searching [32]. The *Chemical Substructure Index* (*CSI*) [33], published by ISI, is a monthly index to *Current Abstracts of Chemistry and Index Chemicus* (*CAC&IC*), and enables subscribers to locate by means of substructure those compounds newly reported in the literature and indexed in *CAC&IC*. This is achieved through a rotated listing of WLNs, in which each WLN is indexed an average of four times.

Machine-readable WLN files can be searched using techniques similar to those employed in nomenclature searching, including the use of symbol proximity and string truncation operators [34]. In common with nomenclature files, a rapid screening search is conducted prior to WLN string search. In the case of ISI's RADIICAL (Retrieval and Automatic Dissemination of Information from Index Chemicus and Line Notations) substructure search program, the screens consist of notation symbols and substrings generated directly from the WLN [35]. In CROSSBOW (Computer Retrieval of Organic Structures Based on Wiswesser), screens are generated from the WLN and from a connection table generated specifically for this purpose. Here, screens are selected to describe features of the structure which are poorly catered for by string search. Thus the notation substrings *QVR* and space-letter-*VQ* give rise in CROSSBOW to a 'substituted carboxylic acid' screen, for which bits 69 and 70 in a 148-bit string indicate single and multiple occurrence respectively.

In common with nomenclature fragments, the correspondence between WLN symbols and substrings and substructures themselves depends upon their context within the structure. As a result, framing a substructure query for string searching can be difficult, and may result in poor precision and the failure to retrieve all relevant answers. In addition to possible errors in the coding of the query and omission of significant strings as search parameters, limitations exist which are inherent in WLN itself. For example, the method of citing cyclic structures in WLN can result in a simple ring being obscured when embedded in a more complex ring system. String searching for specific rings, for example pyridines or quinolines, is relatively easy, but searching for an embedded aromatic, six-membered, N-heterocycle is difficult without enumerating specific examples of ring systems in the query statement. A discussion of some general aspects of the substructure search of the *Index Chemicus Registry System* (*ICRS*) [36], the machine-readable equivalent of *CAC&IC* is available elsewhere [37].

In many cases, the combination of WLN fragment and string searching is sufficient to provide adequate retrieval performance from notation files for queries comprising functional groups and specified ring systems. Greater search specificity is made possible for a wider range of substructural types by converting the WLN into a connection table, from which additional screens may be generated as noted above, and to which atom-by-atom matching algorithms can be applied. In the case of CROSSBOW, the screens which may be assigned from the CROSSBOW connection table fall into several categories. Unit symbol screens correspond to the unit symbols of the connection table, but are of only limited value as the earlier screen and string searches will have ensured the presence or absence of the less common units. Ring screens specify the presence or absence of rings, fusions, bridges, spiro-links and peri-fusions, and the connectivity and size of individual rings. The CROSSBOW *network search* consists of a symbol-by-symbol match of sequences of up to 10 unit symbols from the query substructure against the connection tables of file compounds which pass the unit and ring screens. The network search algorithm operates on each candidate structure by successively ignoring those symbols defined as optional in the query until either a network match is found or the structure is rejected.

Aspects of CROSSBOW searching have been described here in some detail because they illustrate a range of screening techniques, including the use of screens generated automatically from a notation and from a connection table. Of the structure retrieval systems reviewed in Chapter 7, CROSSBOW is the only one to use WLN as the principal structure representation. The CROSSBOW connection table is unique to this system, and is not typical of the connection tables used in other systems (see Chapter 5). Furthermore, CROSSBOW at present lacks the graphics interface for query structure input now used in systems such as CAS ONLINE, Télésystèmes-DARC and MACCS. In these systems, the query structure may be drawn on the terminal screen and a cor ction table is created automatically. Screens are generated from the connectic table by simple computer algorithms, and screening and atom-by-atom s rches proceed wholly automatically. Graphics structure input facilitates the se of these systems by the chemist, to whom WLN remains unfamiliar and tl need for query encoding is irksome.

6.5 SUBSTRUCTURE SEARCH OF CONNECTION TABLES

Many of the problems associated with substructure search of chemical nomenclature and notation files may be overcome in structure retrieval systems based on connection tables. Here, queries are not bound by nomenclature fragments or notation substrings but can be expressed graphically. Furthermore the explicitness of the connection table leads in principle to flexible structure searching characterized by both complete recall and 100 per cent precision.

However, the explicit nature of the connection table means that every substructure must be searched for expressly, and the economy and consistency of performance of substructure searches of connection table files is achieved only by virtue of an effective screening system. Accordingly, a number of screens for connection table files include fragments centred upon atoms and a variety of screen types generated automatically from the connection table. In addition to the molecular formula, element counts and ring descriptors, screens for connection table files include fragments centred upon atoms and bonds of the structure and which describe structural environments of limited size, and linear sequences of atoms and bonds which describe connected paths through the structure. Indeed, as noted earlier, the variety of screens which can be generated automatically from a connection table is unlimited, and hence strict criteria are required to determine the screens to be used in a screening system.

6.5.1 CAS ONLINE Substructure Search System

For the development of CAS ONLINE, CAS chose to use a screen set based upon one developed by BASIC [38], which in turn had been developed from an earlier CAS experimental system and influenced by the results of research conducted at that time by Adamson et al. [8, 13]. However, unlike a screen set selected on the basis of extensive statistical analysis of a structure file, the initial CAS screens had been developed on an empirical basis and, additions notwithstanding, both the CAS ONLINE and the present BASIC screen sets [39, 40] retain this empirical basis.

CAS ONLINE uses a range of twelve screen types, and these are listed in Table 6.3.

Table 6.3 – CAS ONLINE screen types.

Augmented atom screens		Linear sequence screens		General structural feature screens	
AA	Augmented atom	AS	Atom sequence	RC	Ring count
HA	Hydrogen augmented atom	BS	Bond sequence	TR	Type of ring
TW	Twin augmented atom	CS	Connectivity sequence	AC	Atom count
				DC	Degree of connectivity
				EC	Element composition
				GM	Graph modifier

Augmented atoms (*AA*) are atom-centred screens and describe atoms and each of their non-hydrogen attachments, together with the bonds which connect them. Partial descriptions of each *AA* screen are also generated by omitting bond orders, and by omitting attached atoms. As in the case of more generic screens, partial descriptions of *AA* screens facilitate the encoding of substructure queries since these allow for the retrieval of larger structures which contain the desired substructure. *Hydrogen augmented atoms* (*HA*) allow the exact specification of the number of hydrogen atoms attached to the central atom, while *Twin augmented atoms* (*TW*) further specify the number of hydrogens of one of the attached atoms. These screens become useful where it is necessary to specify exactly the degree of substitution permitted on one or more atoms of the query, and are necessary since *AA* screens that do not specify fully the attachments to the central atom imply that either hydrogen or non-hydrogen atoms can be present to complete the valence requirements.

Atom sequence (*AS*) screens are paths of connected atoms of length 4, 5 and 6 non-hydrogen atoms. These are selected as screens for CAS ONLINE according to their ability, in combination, to describe patterns of chemical significance, for example disubstitution patterns on a carbocycle. Bond types are specified in the more common *AS* screens in order to increase their specificity, but bond orders are not cited. Conversely, *Bond sequence* (*BS*) screens of lengths 3, 4 and 5 bonds always specify bond types and often bond orders, but the identity of each atom in a *BS* screen is ignored. *Connectivity sequences* (*CS*) describe the non-hydrogen connectivity of each atom in connected paths of lengths 3, 4, 5 and 6 atoms, and may include bond types but not bond orders. Unlike the *AA*, *AS* and *BS* screens which may be used as partial descriptions of larger, more highly connected environments, *CS* screens specify exactly the connectivity of each atom. Consequently, the value of *CS* screens for substructure search is somewhat limited relative to atom-sequence and bond-sequence screens.

Ring count (*RC*) screens specify the number of rings present in a structure. If a structure contains four rings, for example, the *RC* screens which specify the presence of one, two, three and four rings will be assigned. The node sequences of rings of between 3 and 7 atoms in size are described by *Type of ring* (*TR*) screens. A single *TR* screen is available to indicate the presence of an 8-membered or larger ring. *Atom count* (*AC*) and *Element composition* (*EC*) screens specify the number of non-hydrogen atoms present and the number of atoms of each element respectively. *Degree of connectivity* (*DC*) screens specify the number of atoms which have at least the specified number of non-hydrogen attachments. As in the case of *RC* screens, the assignment of an *AC*, *EC* and *DC* screen results also in the assignment of screens which specify fewer occurrences of the particular characteristic described. *Graph modifier* (*GM*) screens have a more general purpose than the screens described above. *GM* screens are used to identify chemical substance classes, for example alloys, incompletely defined substances, mixtures, polymers and radical ions, and to describe unusual structural features, for example abnormal mass, valence or charge attributes of atoms in the structure [19].

Almost 40 per cent of the screens in the CAS ONLINE screen set occur in 1 per cent or fewer of the structures in the *CA* Registry file, and more than 70 per cent occur in 3 per cent or fewer structures. This distribution, although far from the theoretical ideal, represents an acceptable balance between the theoretical considerations and the need to provide a sufficiently large number of screens of both a general and a specific nature for screening a large and heterogeneous structure file. The inclusion of very general screens can lead to an unfavourable increase in the imbalance of this distribution, but this can be partly overcome by specifying the number of occurrences of these screens in the structure. The count associated with each screen is not an absolute value, but is such that, if assigned to a structure, all screens which specify a lesser occurrence of that same fragment or characteristic are also assigned.

The frequency of assignment of the more highly specific screens can be increased by sharing a single screen number between sets of these screens. As in the case of a superimposed code, some loss in selectivity might be expected as a result of the OR logic implicit among shared screens. However, no significant loss is observed in practice since the dominant influence on screening performace is the number of screens related by AND logic. The major practical benefit of screen sharing is the increase in the number of screens available for substructure searching without a consequent increase in the length of the bit string for each structure. Accordingly, almost 6000 different screens are used in CAS ONLINE, and are represented in a 2128-bit string. Some screen sharing is implicit in the screens themselves, for example where a screen uses a generic symbol to describe a set of related elements, such as 'X' to represent the halogen atoms. Other screens, notably linear sequence screens, which differ only in one atom or in the pattern of bonds, may also be grouped together under one screen number. For example, *AS* screens which differ only in the presence of a nitrogen, oxygen or sulphur atom at one position in the atom sequence frequently share a single screen number.

The CAS ONLINE screens are small in size, typically describing patterns of between 3 and 6 atoms and bonds. As a result of the small screen size and of the large number of general screens and the extent of screen number sharing among the more specific screens, the average frequency of assignment tends to be high. For this reason a serial search file organization is favoured. However, a practical problem arises when searching very large files since, as noted earlier, the entire file must be searched, and search times can be expected to increase with the size of the file. A novel hardware configuration has been developed for CAS ONLINE, and consists of a network of pairs of minicomputers over which the search task is distributed. The pairs of minicomputers operate independently of one another, each searching only a portion of the file. The elapsed time for the screen search depends only on the size of the largest file segment searched by a single minicomputer. Average search times can be set at any desired level simply by choosing the appropriate size for the file segments and adding further minicomputers as appropriate. At present each minicomputer searches a file segment of the *CA* Registry file of some 750,000 structures.

Atom-by-atom searches in CAS ONLINE are conducted on the second minicomputer of each pair and candidate structures are distributed among these as they become identified by the screen search. Flexible resource management ensures the optimum allocation of atom-by-atom searches among these machines, and additional minicomputers can be switched to this task if necessary. The CAS ONLINE system architecture is shown schematically in Fig. 6.3. This architecture has been adapted for use in the Upjohn COUSIN Compound Information System; other aspects of both CAS ONLINE and COUSIN are described in Chapter 7.

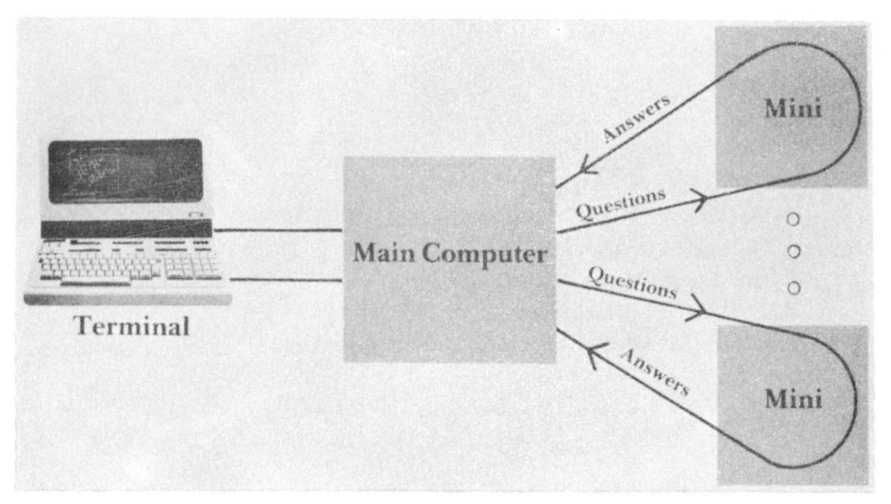

Fig. 6.3 – CAS ONLINE system architecture. (Copyright American Chemical Society, reprinted by permission).

6.6 STRUCTURE SEARCH OF GENERIC CHEMICAL STRUCTURES

The success of substructure search systems for searching files of specific chemical structures based on connection tables, and in particular of commercial services such as CAS ONLINE, has highlighted the need for improved facilities for searching files of generic chemical structures [41]. Generic structures, or *Markush* formulae, are used frequently in chemical patents as the means of expressing a class of related compounds for which the granting of patent rights is claimed. Fig. 6.4 illustrates a typical generic structure from a UK patent. Generic chemical structures are characterized by the following features:

(i) the variable nature of substituent groups, usually expressed as a list of discrete alternative members,
(ii) variable substitution patterns which specify the possible positions of attachment of these groups,
(iii) the use of generic as well as specific nomenclature to express the chemical nature of substituent groups, and
(iv) logical dependencies or exclusions which constrain the particular combinations of these groups.

Patent information services such as those provided by Derwent Publications enable industry to keep abreast of the commercial market and the activities of their competitors worldwide. The use of fragmentation codes for encoding, storing and searching generic structural descriptions has been described in Chapter 5. Several groups have been working independently towards systems which are based on a connection table representation for generic structures, for

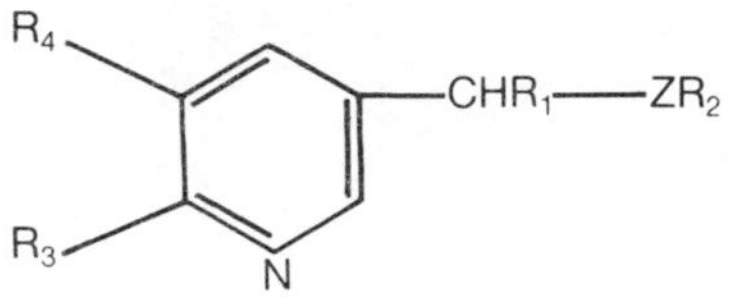

R_1 and R_2 are H, opt. substd. hydrocarbyl, aryloxy or alkoxy, and R_2 may also be pyridyl, thienyl or furyl;
R_3 and R_4 are H, alkyl or aryl, or $R_3 + R_4$ form a bridging group;
Z is CO or $C(OH)R_5$; and R_5 is H, alkyl or aryl.
Provided that when R_1 is H, R_2 is not 4-ClC_6H_4.

Fig. 6.4 – Markush claim from UK patent GB 2015-524.

which systems for graphic structure input and display can be adapted and used, and to which methods of substructure search described in this section can be applied [42–44].

Lynch's group [45] has reported the development of a structure description language GENSAL for the encoding, machine input and display of generic structures from patents. A connection table representation has also been described which records exactly the variable topology and chemical nature of a wide variety of types of generic structure. A number of substructure search techniques are being investigated by this group, among which is a screening algorithm based on the use of CAS ONLINE screens, and a graph-matching algorithm similar to the set reduction algorithms applied to specific chemical structures.

Should these or similar developments prove to be practicable, in terms of both search performance and the economics of establishing and maintaining such systems, the investment by patent information services in upgrading existing generic structure retrieval systems is likely to be substantial, as dictated by the needs of industry for improved access to chemical structure information in patents.

REFERENCES

[1] Almond, J. R., and Welsh, H. M., Chemical Substructure Searching: Industrial Applications and Commercial Systems, *Drexel Library Quarterly*, 84–105 (1982).

[2] Read, R. C., and Corneil, D. G., The Graph Isomorphism Disease, *J. Graph Theory*, **1**, 339–363 (1977).

[3] Ash, J. E., and Hyde, E., (eds.) *Chemical Information Systems*, Ellis Horwood, Chichester (1975).

[4] Sussenguth, E. H., A Graph-Theoretic Algorithm for Matching Chemical Structures, *J. Chem. Doc.*, **15**, 36–43 (1965).

[5] Figueras, J., Substructure Search by Set Reduction, *J. Chem. Doc.*, **12**, 237–244 (1972).

[6] Dubois, J. E., DARC System in Chemistry, in: *Computer Representation and Manipulation of Chemical Information*, Wipke, W. T., Heller, S., Feldmann, R., and Hyde, E. (eds.), Wiley, New York (1974).

[7] Adamson, G. W., Clinch, V. A., and Lynch, M. F., Relationship between Query and Database Microstructure in General Substructure Search Systems, *J. Chem. Doc.*, **13**, 133–136 (1973).

[8] Adamson, G. W., Bush, J. A., McLure, A. H. W., and Lynch, M. F., An Evaluation of a Substructure Search Screen System Based on Bond-Centred Fragments, *J. Chem. Doc.*, **14**, 44–48 (1974).

[9] Feldman, A., and Hodes, L., Substructure Search with Queries of Varying Specificity, *J. Chem. Inf. Comput. Sci.*, **19**, 125–129 (1979).

[10] Farmer, N. A., and O'Hara, M. P., CAS ONLINE. A New Source of Substance Information from Chemical Abstracts Service, *Database*, **3**, 10–25 (1980).

[11] Mooers, C. N., Zatocoding Applied to Mechanical Organization of Knowledge, *Amer. Doc.*, **2**, 20–32 (1951).

[12] Lynch, M. F., Screening Large Chemical Files, in ref [3] Ch. 12.

[13] Adamson, G. W., Cowell, J., Lynch, M. F., McLure, A. H. W., Town, W. G., and Yapp, M., Strategic Considerations in the Design of a Screening System for Substructure Searches of Chemical Structure Files, *J. Chem. Doc.*, **13**, 153–157 (1973).

[14] Lynch, M. F., Variety Generation – A Reinterpretation of Shannon's Mathematical Theory of Communication, and its Implications for Information Science, *J. Amer. Soc. Inf. Sci.*, **28**, 19–25 (1977).

[15] Adamson, G. W., Lynch, M. F., and Town, W. G., Analysis of Structural Characteristics of Chemical Compounds in a Large Computer-Based File. Part 2. Atom-Centred Fragments, *J. Chem. Soc.* (C), 3702–3706 (1971).

[16] Willett, P., A Screen Set Generation Algorithm, *J. Chem. Inf. Comput. Sci.*, **19**, 159–162 (1979).

[17] Feldman, R. J., Milne, G. W. A., Heller, S. R., Fein, A., Miller, J. A., and Koch B., An Interactive Substructure Search System, *J. Chem. Inf. Comput. Sci.*, **17**, 157–163 (1977).

[18] Graf, W., Kaindl, H. K., Kniess, H., Schmidt, B., and Warszawski, R., Substructure Retrieval by Means of the BASIC Fragment Search Dictionary Based on the Chemical Abstracts Service Chemical Registry III System, *J. Chem. Inf. Comput. Sci.*, **19**, 51–55 (1979).

[19] Dittmar, P. G., Farmer, N. A., Fisanick, W., Haines, R. C., and Mockus, J., The CAS ONLINE Search System. I. General System Design and Selection, Generation, and Use of Search Screens, *J. Chem. Inf. Comput. Sci.*, **23**, 93–102 (1983).

[20] Feldman, A., and Hodes, L., An Efficient Design for Chemical Structure Searching, 1. The Screens, *J. Chem. Inf. Comput. Sci.*, **15**, 147–152 (1975).

[21] Page, J., Theisen, R., and Kuhl, K., The Walter Reed Army Institute of Research Chemical Information System, *ACS Symposium Series*, **84** (1978).

[22] Hodes, L., Selection of Descriptors According to Discrimination and Redundancy. Application to Chemical Structure Searching, *J. Chem. Inf. Comput. Sci.*, **16**, 88–93 (1976).

[23] Meyer, E., Superimposed Screens for the GREMAS System, in: *Mechanized Information, Storage, Retrieval and Dissemination*, North-Holland, Amsterdam (1968).

[24] Hodes, L., A Square Root Algorithm for Inclusive Matching. Application to Chemical Structure Searching, *Proceedings of the Conference on Computer Graphics, Pattern Recognition, and Data Structures*, IEEE Computer Society (1975).

[25] Callahan, M. V., and Rusch, P. F., Online Implementation of the CA SEARCH File and CAS Registry Nomenclature File, *Online Review*, **5**, 377–393 (1981).

[26] Pottier, P. E., Substructure Searching in CHEMLINE, *Online*, **1**, 23–25 (1977).

[27] Fisanick, W., Mitchell, L. D., Scott, J. A., and Vander Stouw, G. G., Substance Searching of Computer-readable Chemical Abstracts Service Ninth Collective Index Chemical Nomenclature Files, *J. Chem. Inf. Comput. Sci.*, **15**, 73–84 (1975).

[28] Dunn, R. G., Fisanick, W., and Zamora, A., A Chemical Substructure Search System Based on Chemical Abstracts Index Nomenclature, *J. Chem. Inf. Comput. Sci.*, **17**, 212–219 (1977).

[29] Harrison, M. D., Implementation of the Substring Test by Hashing, *Commun. Amer. Comput. Mach.*, **14**, 777–779 (1971).

[30] Hartwell, I. O., Using the New Features in the DIALOG Chemical Information System, *Database*, 11–23 (1983).

[31] Rowland, J. F. B., and Veal, M. A., Structure-Text and Nomenclature-Text Searching for Chemical Information: An Experiment with the Chemical Abstracts Integrated Subject File and Registry System, *J. Chem. Inf. Comput. Sci.*, **17**, 81–89 (1977).

[32] Granito, C. E., Schultz, J. E., Gibson, G. W., Gelberg, A., Williams, R. J., and Metcalf, E. A., Rapid Structure Searches via Permuted Chemical Line Notations. III. A Computer-Produced Index, *J. Chem. Doc.*, **5**, 229–233 (1965).

[33] Granito, C. E., and Rosenberg, M. D., Chemical Substructure Index (CSI). A New Research Tool, *J. Chem. Doc.*, **11**, 251–256 (1971).

[34] Crowe, J. E., Leggate, P., Rossiter, B. N., and Rowland, J. F. B., The Searching of Wiswesser Line Notations by Means of a Character-matching Serial Search, *J. Chem. Doc.*, **13**, 85–92 (1973).

[35] Granito, C. E., Becker, G. T., Roberts, S., Wiswesser, W. J., and Windlinx, K. J., Computer-Generated Substructure Codes (Bit Screens), *J. Chem. Doc.*, **11**, 106–110 (1971).

[36] Garfield, E., Revesz, G. S., Granito, C. E., Dorr, H. A., Calderon, M. M., and Warner, A., Index Chemicus Registry System: Pragmatic Approach to Substructure Chemical Retrieval, *J. Chem. Doc.*, **10**, 54–58 (1970).

[37] Leggate, P., Rossiter, B. N., and Rowland, J. F. B., Evaluation of an SDI Service Based on the Index Chemicus Registry System, *J. Chem. Doc.*, **13**, 192–203 (1973).

[38] Schenk, H. R., and Wegmuller, F., Substructure Search by Means of the Chemical Abstracts Service Chemical Registry II System, *J. Chem. Inf. Comput. Sci.*, **16**, 153–161 (1976).

[39] *CAS ONLINE Screen Dictionary for Substructure Search*, 2nd edn., Chemical Abstracts Service, Columbus, Ohio (1981).

[40] Graf, W., Kaindl, H. K., Kniess, H., and Warszawski, R., The Third BASIC Fragment Dictionary, *J. Chem. Inf. Comput. Sci.*, **22**, 177–181 (1982).

[41] Silk, J. A., Present and Future Prospects for Structural Searching of the Journal and Patent Literature, *J. Chem. Inf. Comput. Sci.*, **19**, 195–198 (1979).

[42] Lynch, M. F., Barnard, J. M., and Welford, S. M., Computer Storage and Retrieval of Generic Chemical Structures in Patents. Part 1. Introduction and General Strategy, *J. Chem. Inf. Comput. Sci.*, **21**, 148–150 (1981).

[43] Nakayama, T., and Fujiwara, Y., Computer Representation of Generic Chemical Structures by an Extended Block-Cutpoint Tree, *J. Chem. Inf. Comput. Sci.*, **23**, 80–87 (1983).

[44] Kudo, Y., and Chihara, H., Chemical Substance Retrieval System for Searching Generic Representations. 1. A Prototype System for the Gazetted List of Existing Chemical Substances of Japan, *J. Chem. Inf. Comput. Sci.*, **23**, 109–117 (1983).

[45] Welford, S. M., Lynch, M. F., and Barnard, J. M., Towards Simplified Access to Chemical Structure Information in the Patent Literature, *J. Inf. Sci.*, **6**, 3–10 (1983).

7

Chemical structure search systems and services

The previous chapters have described the various forms of structure representation suited to computer manipulation, and the techniques used for searching structure records for specific and generalized structural information. In this chapter a number of chemical structure storage and retrieval software systems are reviewed. These are listed in Table 7.1.

Table 7.1 – Chemical structure search systems.

Acronym	Search system
CAS ONLINE	Chemical Abstracts Service. Online Substructure Search Service.
CIS SANSS	National Institute of Health/Environmental Protection Agency (NIH/EPA). Chemical Information System: Structure and Nomenclature Search System.
COUSIN	Upjohn Company. Compound Search Information System.
CROSSBOW	Imperial Chemicals Industry (ICI). Computerized Retrieval of Organic Structures Based on Wiswesser.
DARC	Télésystèmes-Questel. Description, Acquisition, Retrieval and Correlation.
MACCS	Molecular Design Limited Inc. (MDL). Molecular Access System.

The development of these systems spans a period of some twenty years from the earliest implementation of CROSSBOW in the 1960s to the introduction of the CAS ONLINE and Télésystèmes–DARC services in the early 1980s. Certain of the systems are available as 'off the shelf' software packages and can be purchased for use with company databases. Others are used for searching commercial databases and are accessed through national and international telecommunications networks. Price information and specific technical details of each system are subject to continuous change and are not given here. These and further details may be obtained from the references listed at the end of the chapter, and from the organization responsible for the development or marketing of the system, for which purpose a list of addresses is included at the end of the book (Appendix 2). Notable software systems which are not reviewed here include CHEMPIX, a system developed at Roussel-Uclaf and marketed by Chemical Information Management Inc. (CIMI), and the SOCRATES system developed at Pfizer Central Research in the UK.

Section 7.1 describes the history of the development of chemical structure search systems and services, while section 7.2 provides short summaries of the systems listed in Table 7.1. Section 7.3 describes aspects of each system relating to the methods of structure representation, search file organization and substructure search, and highlights certain of the points made in the preceding chapters. System developments since the late 1970s have been concerned largely with enhancements to the access and use of each system, rather than with dramatic changes in the basic technologies involved. This trend is not without exception, for example the distributed hardware search machine adopted for the CAS ONLINE service which marks a significant advance in the design of computer systems for searching very large databases. One area in which recent enhancements have been most striking is in the methods of query structure input and structure display, for which graphics software and hardware devices are described in Chapter 10. Accordingly, the methods of structure input and display for the system reviewed here are described in section 7.4.

7.1 IN-HOUSE AND COMMERCIAL SERVICES

Recent years have seen a considerable change in the role of the information departments of chemical industries, and particularly in the research-based pharmaceuticals and agrochemicals companies. Traditionally this role has consisted primarily of the gathering and indexing of published documents, including journal articles, reports and patents, of particular relevance to the company's interests, and providing current awareness and reference services to individuals and to departments within the company. In addition, the information departments have been responsible for the gathering and organizing, for use by the company, of data originating from within the company itself, the *in-house* database. This latter responsibility has remained essentially unchanged, and in the case of the larger organizations the provision of services from the in-house database has been augmented with a variety of new computer-based services. These include the use of software packages for the purposes of structure modelling and for the elucidation of structural information from analytical data, and for quantitative structure–activity correlation and the identification of synthetic pathways. Examples of these applications are described in Chapter 9.

Rising costs and unnecessary duplication of effort have led to the decline of decentralized abstracting and indexing of the published literature and, whilst current awareness bulletins are still prepared and distributed locally, retrospective searching of the published chemical literature is now provided almost exclusively through the products of the major commercial documentation services, notably Chemical Abstracts Service (CAS), Derwent Publications, and the Institute for Scientific Information (ISI). During this period of transition a number of collaborative schemes have become established between otherwise competing organizations with the purpose of reducing costs. In these groups the task of information gathering and of database creation and maintenance is shared among the participating organizations, and information services are provided centrally to each member of the group. Examples of these groups include the Pharma Dokumentationsring (PDR) [1] and the Internationale Dokumentationsgesellschaft für Chemie (IDC) [2,3] in Germany, and the Basel Information Centre (BASIC) [4] in Switzerland. In each of these cases, the collaborative efforts have led to services within the group which were not available elsewhere, and which have since formed the basis of commercial services to the whole community. For example, the RINGDOC service established by PDR for indexing and searching specific chemical substances was subsequently taken over by Derwent Publications and offered as a commercial search service. Similarly, the substructure search system developed by BASIC from an early CAS system now forms the basis of the present CAS ONLINE service.

The Pharma Dokumentationsring was established in 1958, and indexes relevant information from selected publications, using the fragmentation code RINGCODE [5] to encode structural information on specific compounds. As a result of dissatisfaction with the results obtained using the early Derwent patent documentation services [6], PDR proceeded to encode chemical patents using RINGCODE, including generic structural information for which an automatic encoding program CORA was developed [7]. PDR considered it necessary to encode patents in this way up to 1977, since when patent encoding in RINGCODE has ceased both for reasons of economy and because improvements in the Derwent CPI (Central Patents Index) code rendered this recoding largely unnecessary [8].

The Internationale Dokumentationsgesellschaft für Chemie was formed in 1967 and is now supported by eleven member organizations, ten of which are in Germany and one in Japan. Until 1974, IDC indexed journal articles of particular relevance to synthetic chemistry selected by chemists from the published literature. Chemical structure information was encoded into the GREMAS fragmentation code [9], and information abstracts were prepared manually. Since that time suitable articles have been selected by means of the online search file CA SEARCH and, where appropriate, additional indexing information and augmented abstracts are added. Encoding of specific compounds in GREMAS is performed automatically by reprocessing structure records from the CAS Chemical Registry system, tapes of which are supplied to IDC under licence. IDC continues to encode Markush structures from chemical patents in the GREMAS code, in most cases using Derwent CPI abstracts as source material and consulting the original patent specifications only when necessary. The role of IDC has changed significantly since its inception, and now forms a part of the German National Information Centre for Chemistry and as such offers a public search service of its combined patent and non-patent files.

The Basel Information Centre provides a similar service to its member companies by reprocessing bibliographic and structure files supplied under licence from CAS since 1968. The experience gained at BASIC in the development and use of an online substructure search system [10, 11] has benefited CAS in the design of the CAS ONLINE service [12, 13]. Indeed, the advent of substructure search services such as CAS ONLINE and Télésystèmes-DARC [14–19], together with the NIH-EPA Chemical Information System [20–24], has increased greatly the ability to search retrospectively the numerous chemical databases and databanks now available, and fully realizes the enormous volume and variety of information available to organizations at only modest cost.

The rather slow development of these commercial services contrasts sharply with the development of information systems within industry, and particularly within research-based organizations. Here the pace of innovation and the adoption of new technologies have been more rapid, arising from the need to provide and maintain current and effective services within each organization. Although commercial services are now able to offer remote storage facilities for company databases and to search these using their own search software, this option is seldom chosen for reasons of confidentiality and security. Instead, and often at considerable expense, companies have developed computer-based systems to be used specifically with their own in-house databases. However, the spiralling cost of software development now leads most companies to purchase for in-house use a structure storage and retrieval system in the form of a software package, which includes installation and database conversion, software maintenance and periodic upgrading. Of the systems reviewed in this chapter, some have been developed within pharmaceutical companies (CROSSBOW [25–28], COUSIN [29–31]), one within a university (Télésystèmes-DARC [14–17]), one in a government agency (NIH-EPA CIS [20–24]), and one within an existing major documentation service (CAS ONLINE [12, 13]). Specialist software houses have also arisen to cater for this market, and the MACCS system [32, 33] of Molecular Design Limited Inc. is the most successful of the products from this source.

7.2 SUBSTRUCTURE SEARCH SYSTEMS

7.2.1 CAS ONLINE

CAS ONLINE [12, 13] is perhaps the most important of the available public-access substructure search systems owing to the unique position of CAS among the major chemical documentation services. CAS ONLINE was introduced in 1981 as a dial-up service for online searching of the CAS Registry file, which at present contains searchable records for over 6.5 million chemical substances indexed by CAS since 1965, with a monthly addition of some 35,000 substances. Searching of *Chemical Abstracts* (*CA*) bibliographic and index entry data is also available through CAS ONLINE, in which it is possible to transfer to the bibliographic CA FILE the set of Registry Numbers (RNs) for substances retrieved through substructure search of the Registry file (see Chapter 6). The CAS ONLINE search software is not available for purchase and use in-house, but company databases can be mounted and searched as private registry files on the CAS computers in Columbus, Ohio.

7.2.2 Télésystèmes-DARC

The DARC (Description, Acquisition, Retrieval and Correlation) method of encoding and searching chemical structure information was devised by Dubois at the University of Paris [14–17], and has been further developed through the 1970s by the Association pour la Recherche et le Développement en Informatique Chimique (ARDIC) [18]. The DARC system is now maintained as an online substructure search facility by Télésystèmes-Questel [19]. Under licence with the American Chemical Society (ACS) the DARC software has been used commercially since 1981 to search the CAS Registry file and other CAS-derived structure files. In the EURECAS service available through Télésystèms-Questel, the results of DARC substructure searches can be used to search the CAS bibliographic files and other databanks using the Questel text-search software. Télésystèmes-Questel have recently announced the exclusive availability of online search of the ISI *Current Abstracts of Chemistry and Index Chemicus* (*CAC&IC*), which contains structural and bibliographic data for over 3 million chemical substances. Since late 1981 the DARC software has been marketed by the Centre National de l'Information Chimique (CNIC) for use in-house with company databases, and is at present available for IBM and DEC machines.

7.2.3 NIH-EPA CIS

The NIH-EPA Chemical Information System (CIS) has been in use since the early 1970s and forms a part of a larger US government-sponsored project [20–24]. CIS comprises several major components: the Structure and Nomenclature Search System (SANSS) [23] based on search algorithms developed by Feldmann [20], a large collection of numeric databanks (see Chapter 4), and a number of structure analysis and molecular modelling programs. The CIS software is available from the National Technical Information Service (NTIS) for a nominal fee, for which software documentation and maintenance are not provided. A fully documented and maintained version of SANSS is available from Fein-Marquart Associates, and operates on a DEC computer. Lederle in the US and the Science and Engineering Research Council (SERC) Daresbury laboratory in the UK are established users, though both have invested a substantial amount of local programming effort to augment the software for their own purposes.

7.2.4 CROSSBOW

Since its development at ICI in the 1960s CROSSBOW (Computerized Retrieval of Organic Structures Based on Wiswesser) has become widely established with over two dozen installations worldwide [25–28]. Since 1978 the marketing rights have been held by Fraser Williams (Scientific Systems) Ltd. CROSSBOW is used at ICI to maintain and search an online database of 250,000 compounds, and is also used in batch-mode to search the Fine Chemicals Directory (FCD), the Hansch databases, and the *Index Chemicus Registry System* (*ICRS*) [34], which at present contains some 3 million structure records encoded in the Wiswesser Line-formula Notation (WLN). CROSSBOW also has comprehensive facilities for maintaining and searching property files in addition to structure and substructure search capabilities. CROSSBOW is available for a large variety of machines and has been installed on Burroughs, UNIVAC, FACOM, IBM, DEC and PR1ME computers.

7.2.5 MACCS

MACCS (Molecular Access System) [32, 33] is a modular software package developed and marketed since 1979 by Molecular Design Limited Inc. (MDL). MACCS offers versatile substructure search facilities on databases of up to 400,000 structures, as well as facilities for storing and retrieving supplementary data, such as spectral data and biological test results. MDL also markets software modules for molecular modelling, pattern recognition, structure elucidation and report generation, and a system REACCS for reaction indexing (see Chapter 8). MACCS has rapidly achieved over twenty installations worldwide, and is available for a wide variety of computers, including PR1ME, DEC, Honeywell (MULTICS), IBM and Fujitsu machines. Maintenance and software upgrades are provided, and training and user-support are included in the purchase price.

7.2.6 COUSIN

COUSIN (Compound Information System) has been developed at the Upjohn Company in the USA [29–31], but unlike the CROSSBOW system which was also developed by a pharmaceutical company for its own use, COUSIN is not commercially available as a software package. COUSIN has been in use at Upjohn since 1980 and provides substructure retrieval and access to preclinical data on an in-house database of some 65,000 compounds. COUSIN incorporates many interesting features, of which details may be found in the references cited, and employs a hardware configuration similar to that used for CAS ONLINE and an unreleased IBM relational database package for handling its compound data files. Structure input and search are conducted from DEC GT40 workstations, linked to an IBM 4341. In addition, a DEC PDP 11/55 is dedicated to the atom-by-atom search process [31].

7.3 STRUCTURE REPRESENTATION, ORGANIZATION AND SEARCH

CROSSBOW is the only system to use WLN as the means of canonical structure representation [26]. The bond-implicit CROSSBOW connection table is WLN-based, and differs in this respect from the bond-explicit connection tables used in each of the other systems [35]. CAS ONLINE of course searches CAS Registry III format connection tables [36], which form the basis of each structure record in the present Chemical Registry system. DARC reprocesses the CAS Registry III format connection tables supplied by CAS into the DARÇ structure code developed by Dubois [14–17]. CIS SANSS uses a slightly less detailed form of the CAS connection table, while MACCS additionally includes stereochemical information, and employs a stereochemically extended Morgan algorithm to generate the so-called 'SEMA names' (see Chapter 5). Similarly, COUSIN uses a variant of the Morgan algorithm to provide unique connection table records from graphic structure input.

CIS SANSS, CAS ONLINE and CROSSBOW handle only two-dimensional structure descriptions, while DARC in-house has limited facilities for specifying stereochemistry for structure searching. In COUSIN, bonds 'in' and 'out' of the plane can be drawn at the terminal but stereochemical descriptors are not used for structure search. MACCS enables information on relative and absolute stereochemistry to be stored and uses this information as parameters for structure retrieval.

With regard to limits on structure size, each system has an upper limit dictated by software and hardware characteristics. In the CAS Chemical Registry system specific substances and the structural units of polymers and mixtures are each limited to 253 non-hydrogen atoms. DARC structure storage and retrieval is similarly limited when applied to CAS-derived structure files. MDL offers various versions of MACCS, with compound size increasing from 96 atoms to 255 atoms by steps of 16. The maximum connectivity of any atom is 5, although upgrading to 8 is to be implemented shortly. The CROSSBOW connection table has a limit of 150 atoms, while CIS SANSS permits a maximum of 130 atoms and a maximum connectivity for any atom of 6, on account of requirements of the original Feldmann structure input and display software [20].

7.3.1 Tautomerism and aromaticity

The registration procedures used at CAS, and those used by MACCS and by CROSSBOW have been described in Chapter 5. Other factors, notably tautomerism and aromaticity are described here. CAS uses normalized bonds to cope with tautomeric and alternating bond structures [37]. Tautomeric and delocalized environments are automatically perceived and normalized by replacing the explicit single and double bonds with special tautomer and alternating bond types, with the migrating hydrogen in a tautomer becoming associated with a group of atoms rather than with just a single atom. The user of CAS ONLINE must be aware of this treatment when describing in terms of user-assigned screens query structures which are potentially tautomeric or which contain alternating bonds. Aromatic 6-membered rings are detected automatically by MACCS, and extension of the software to handle tautomers is under consideration. COUSIN and CIS SANSS have automatic perception of tautomers and aromaticity upon registration, regardless of whether a ring is isolated or is part of a larger fused system, and like CAS ONLINE each distinguishes between ring-tautomeric and chain-tautomeric bonds. CROSSBOW adopts very rigorous indexing conventions and treats tautomers as distinct compounds, each registered specifically. As with all system conventions which relate to structure representation, the searcher must follow these exactly when formulating a query structure.

7.3.2 Structure file conversion

Installation of any system in-house may require conversion of an existing database into a format appropriate to the particular system adopted (see Chapter 5). Where the existing file is not in machine-readable form, manual preparation and input of the file is required. Database creation is expensive and highly subject to error, and demands significant resources even for the creation of small files, although expenditure is likely to be recouped quickly by the more efficient and effective operation of the automated system.

The availability of conversion programs for chemical structure files enables a connection table with structure coordinates to be derived from WLN, so that use of WLN as a means of structure entry for database creation may be preferred. Equally, the same programs will convert an existing machine-readable WLN file into the appropriate connection table file. MDL markets a suite of programs to convert a WLN file into MACCS connection tables. Fraser Williams (Scientific Systems) has the marketing rights to a program DARING developed by the SERC in the UK which converts WLN to CIS connection tables [38]. This can be interfaced with Molecular Design's LAYOUT program which derives structure coordinates from the connection tables. In addition, MDL have programs to convert chemical typewriter input and CAS Registry III connection tables into MACCS connection tables and structure coordinates. Télésystèmes–Questel offers programs for structure file conversions between forms of connection table for use with DARC in-house.

7.3.3 File organization

Registration and database updating are performed online by MACCS, COUSIN and in-house installations of DARC. CROSSBOW is used at ICI for online registration of structures input in the form of WLN, but the CROSSBOW package marketed by Fraser Williams (Scientific Systems) Ltd operates only in batch-mode. (CROSSBOW bit and string search is online at ICI but atom-by-atom search and structure display, for which structure coordinates are calculated from the CROSSBOW connection table, are performed in batch-mode [27]).

Each of these systems, except for DARC and CIS SANSS, uses essentially a serial search file organization. In the case of full structure searching, hashing techniques make search times effectively independent of database size, while substructure search times can be expected to increase as a function of database size. As a result, these systems are not well suited to online operation with files larger than several hundreds of thousands of structures. Both CAS ONLINE and COUSIN have overcome this problem by adopting a distributed hardware configuration which ensures a search time which is independent of database size. In the case of CAS ONLINE the Registry file is distributed over a network of pairs of PDP-11 minicomputers, each pair being responsible for approximately 750,000 structures. One machine of each pair is responsible for screening searches on that portion of the file, while the other performs atom-by-atom searches on candidate structure records identified by the screening search. Searching is carried out simultaneously on each partition under the control of an executive program, which additionally controls user dialogue and the display of search output. The hardware is configured in such a way that additional minicomputers can be added to cater for the increasing size of the Registry file, with no observable degradation in search time. A more detailed description of the CAS ONLINE substructure search algorithms may be found in Chapter 6.

DARC and CIS SANSS each employ a form of inverted file organization. As a result, DARC substructure searching is very fast, while search times with SANSS vary widely on account of a comparatively slow atom-by-atom search algorithm. Substructure searching of serial files tends to be less rapid, although this may not be apparent to the user as hits may be inspected as they are found during execution of the search. Despite longer search times, a serial file organization generally enables immediate registration of new compounds, and more timely updating and easier database maintenance, each of which is an important consideration for an in-house system.

Both CAS ONLINE and Télésystèmes-DARC enable a query structure to be searched first against a dynamic sample file, derived from the full CAS Registry file, and from this gives a projection of the number of structures likely to be retrieved from a search of the full Registry file. The query structure may then be modified as appropriate by generalizing or further specifying certain structural features.

All of the systems described here display the results of structure searches in a variety of formats. In addition to nomenclature, registry numbers, indexing information and where appropriate bibliographic citations, output of structure diagrams in the form illustrated in Fig. 7.1 is possible on suitable graphics terminals [39]. In the case of CROSSBOW, structure diagrams are character-based only and can be displayed on a teletype-compatible text terminal. Both character-based and graphic output options are available with NIH/EPA CIS, while in the remaining systems the display of structure diagrams is possible only on a graphics terminal, and is not available to searchers using a text terminal.

```
REG 64603-91-4                                                         ANS 1
IND Isoxazolo[5,4-c]pyridin-3(2H)-one, 4,5,6,7-tetrahydro- (9CI)
MAT THIP
SYN THIP
SYN 4,5,6,7-Tetrahydroisoxazolo[5,4-c]pyridin-3-ol
FOR C6 H8 N2 O2

                     HN      O   NH
                                   O

REFERENCE  1

AN    CA97(19):162964x
TI    Isoxazolo[5,4-c]pyridines which are GABA-agonists
AU    Krogsgaard-Larsen, Povl
CS    Lundbeck, H., og Co. A/S
PA    Can. CA 1,125,288 A2, 08 Jun 1982, 29 pp. Division of Can. Appl. No.
      305,798.
DT    P
PY    1982
LA    Eng
```

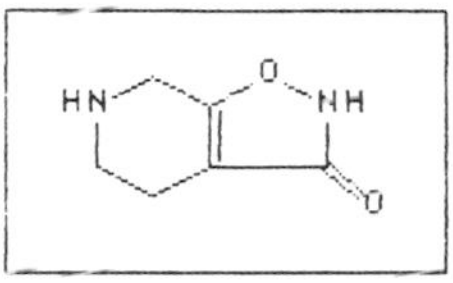

Fig. 7.1 – CAS ONLINE graphic structure display. (CAS ONLINE Guide to Commands. Copyright by the American Chemical Society, reprinted by permission.)

7.4 STRUCTURE INPUT AND DISPLAY

With the exceptions of CROSSBOW and CIS SANSS, each of these packages has available sophisticated structure input and display modules, which allow the chemist to communicate directly with the search system by means of structure diagrams. To make the fullest use of these facilities, vector or raster graphics terminals are required which enable graphic information to be constructed and manipulated on the terminal screen by means of a joystick, light-pen or digitizing tablet (see Chapter 10). In order to provide a permanent record of query structures input, and particularly for graphic structural information output by the search system, a hard-copy thermal printer or plotting device is desirable. In the case of public-access systems, and for those sold widely for in-house use, it is necessary to provide for structure input on less specialized peripherals, specifically the teletype-compatible text terminals used routinely for online access to bibliographic information services.

The CAS ONLINE and Télésystèmes–DARC substructure search services illustrate the options available for interactive query structure input, and each may be used either with text terminals or with graphics devices. In the case of CAS ONLINE, structural fragments may be entered as screen numbers using the CAS ONLINE Screen Dictionary [40] and combined into a query specification using logical operators. Alternatively the query can be constructed at the terminal using either the text structure input (TSI) commands or the menu-driven graphic structure input (GSI) module on an intelligent graphics terminal.

7.4.1 Text structure input

CAS ONLINE text structure input (TSI) permits the construction of a query structure using the commands listed in Table 7.2. It is possible to stack commands, so that the experienced searcher is able to combine all of the structure-building and attribute commands in a single statement [41]. Fig. 7.2 illustrates the use of TSI in constructing a substituted oxazolo-pyridine query structure. In the sequence illustrated, the *5,6* fused ring skeleton is called up in Fig. 7.2(a). Bond values are unspecified and carbon atoms are assumed at all nodes. In Fig. 7.2(b) single node chains are attached at ring positions 4 and 7. Atom types are specified in Fig. 7.2(c) using the NODE command, assigning oxygen at nodes 2 and 10, nitrogen at nodes 3 and 7, and any atom (*A*) at node 11 to indicate any substitution. Bond values are assigned in Fig. 7.2(d) using the BOND command. The command ALL RSE sets all bonds in the structure to Ring Single Exact (*RSE*, symbolized by the hyphen, '-'), while the remaining BOND parameters specify particular bond values at certain locations; specifically a Chain Single Exact (*CSE*) bond between nodes 7 and 11 ('-'), a Ring Double Exact (*RDE*) bond ('=') between fusion nodes 1 and 5, a Ring Unspecified (*RU*) bond ('?') between ring nodes 3 and 4, and a Chain Unspecified (*CU*) bond ('?') to the exocyclic oxygen atom, node 10.

Text structure input in Télésystèmes-DARC follows a similar procedure, but here the structure graph is described by tracing a path through its nodes, numbered arbitrarily. The structure is first defined by a sequence of node numbers whose order describes the connectivity of the atoms of the structure. Discontinuities in the sequence represent branch points and ring closures. Atom and bond values are then assigned to nodes and edges as appropriate. Fig. 7.3 illustrates this form of query structure input. First the graph (*GR*), and then atoms (*AT*), bonds (*BO*) and free sites (*FS*) are specified in response to the system directive -QU-(RN, RQ, CA, GR, BO, AT, FS, CH, VE)?. The fused ring skeleton and its substituents are here defined separately, and heteroatoms are assigned to the appropriate nodes. Aromatic bonds (*AR*) are specified, and then overriden locally by Single (*SI*) and Double (*DO*) bonds between specified nodes. No distinction is made between ring and chain bonds in the DARC structure record, and this is the major cause of some lack of precision in DARC structure retrieval. Free sites, that is positions of permitted but undefined substitution, are specified as desired.

Table 7.2 – CAS ONLINE text structure input: TSTRUCTURE subcommands.

STRUCTURE BUILDING SUBCOMMANDS		
1.	GRAPH (GRA)	to create the structure skeleton assuming unspecified bond values with carbon atoms at all nodes
2.	NODE (NOD)	to specify atoms in a previously built structure
3.	BOND (BON)	to specify bond types and bond values
ATTRIBUTE SUBCOMMANDS		
4.	CHARGE (CHA)	to specify a charge on a node or nodes
5.	MASS (MAS)	to specify mass on a node or nodes
6.	VALENCE (VAL)	to specify valence on a node or nodes
7.	HCOUNT (HCO)	to specify the number of hydrogens attached to a node or nodes
8.	HMASS (HMA)	to specify an abnormal mass (^1H = normal) for hydrogens attached to a node
9.	DLOC (DLO)	to specify a delocalized charge on a set of nodes
10.	NSPEC (NSP)	to specify whether a node is in a ring, a chain, or either
11.	RSPEC (RSP)	to specify whether all rings *must* be isolated or whether they may be isolated or embedded
OTHER SUBCOMMANDS		
12.	DELETE (DEL)	to delete the entire structure of individual nodes, bonds, or attributes
13.	DISPLAY (DIS)	to display the structure image, structure attributes, connection table or combinations or subsets of these items
14.	RESTORE (RES)	to return the structure to the condition before the last set of commands that modified it
15.	GET	to recall a structure from the user's query definition file (without clearing it from that file) so that it can be used in building a new structure
16.	END	to save the structure just completed and terminate the TSTRUCTURE command
17.	HELP	to supply help to the user for each TSTRUCTURE subcommand

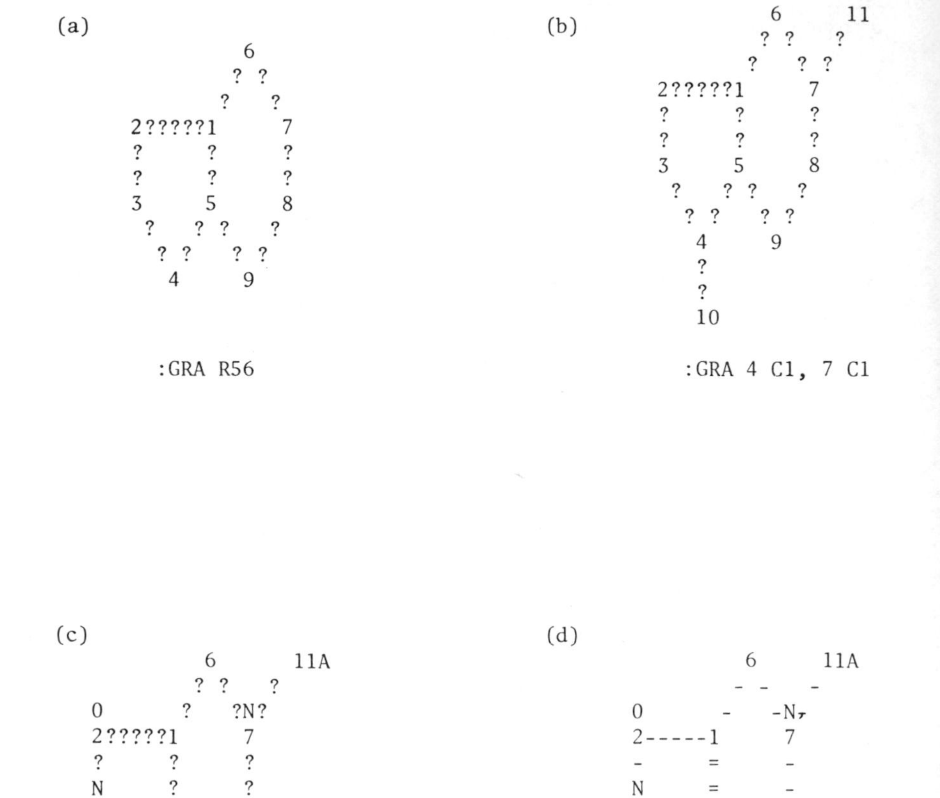

Fig. 7.2 – CAS ONLINE Text Structure Input (TSI). (CAS ONLINE Guide to commands. Copyright by the American Chemical Society, reprinted by permission.)

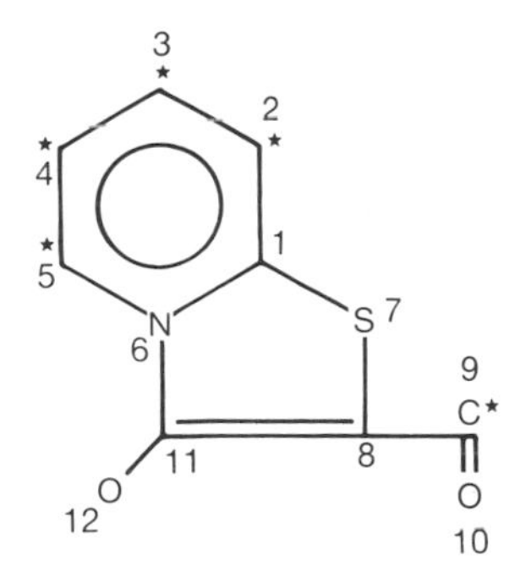


```
-QU-(RN, RQ, CA, GR, BO, AT, FS, CH, VE)? GR
? 1-2-3-4-5-6-1-7-8-11-6
? 11-12, 8-9-10
-QU-(RN, RQ, CA, GR, BO, AT, FS, CH, VE)? AT
? C
? N 6
? S 7
? O 10, 12
-QU-(RN, RQ, CA, GR, BO, AT, FS, CH, VE)? BO
? AR
? SI 1-7-8-9, 12-11-6
? DO 8-11, 9-10
-QU-(RN, RQ, CA, GR, BO, AT, FS, CH, VE)? FS
? 1, 2, 3, 4, 5, 9
```

Fig. 7.3 – Télésystèmes–DARC text structure input. (Reprinted with permission from Centre National de l'Information Chimique.)

7.4.2 Graphic structure input

Menu-driven structure input is common to the graphic input modules in all of these systems. Here, a hierarchy of commands is presented to the user, and appropriate selections lead to subcommands which enable common structural templates to be recalled, displayed and manipulated as desired to form valid structure diagrams. Fig. 7.4 illustrates the command menu in CAS ONLINE graphic structure input (GSI), and the construction of the oxazolo-pyridine structure illustrated previously in Fig. 7.2. The searcher creates a structure by pointing in any order to appropriate menu items and then to positions on the terminal screen using either the terminal cursor buttons or a digitizing tablet and stylus device. A six-membered, normalized ring is displayed by pointing to RING6(N) in Fig. 7.4(a). A pentane ring is then selected and positioned in Fig. 7.4(b)

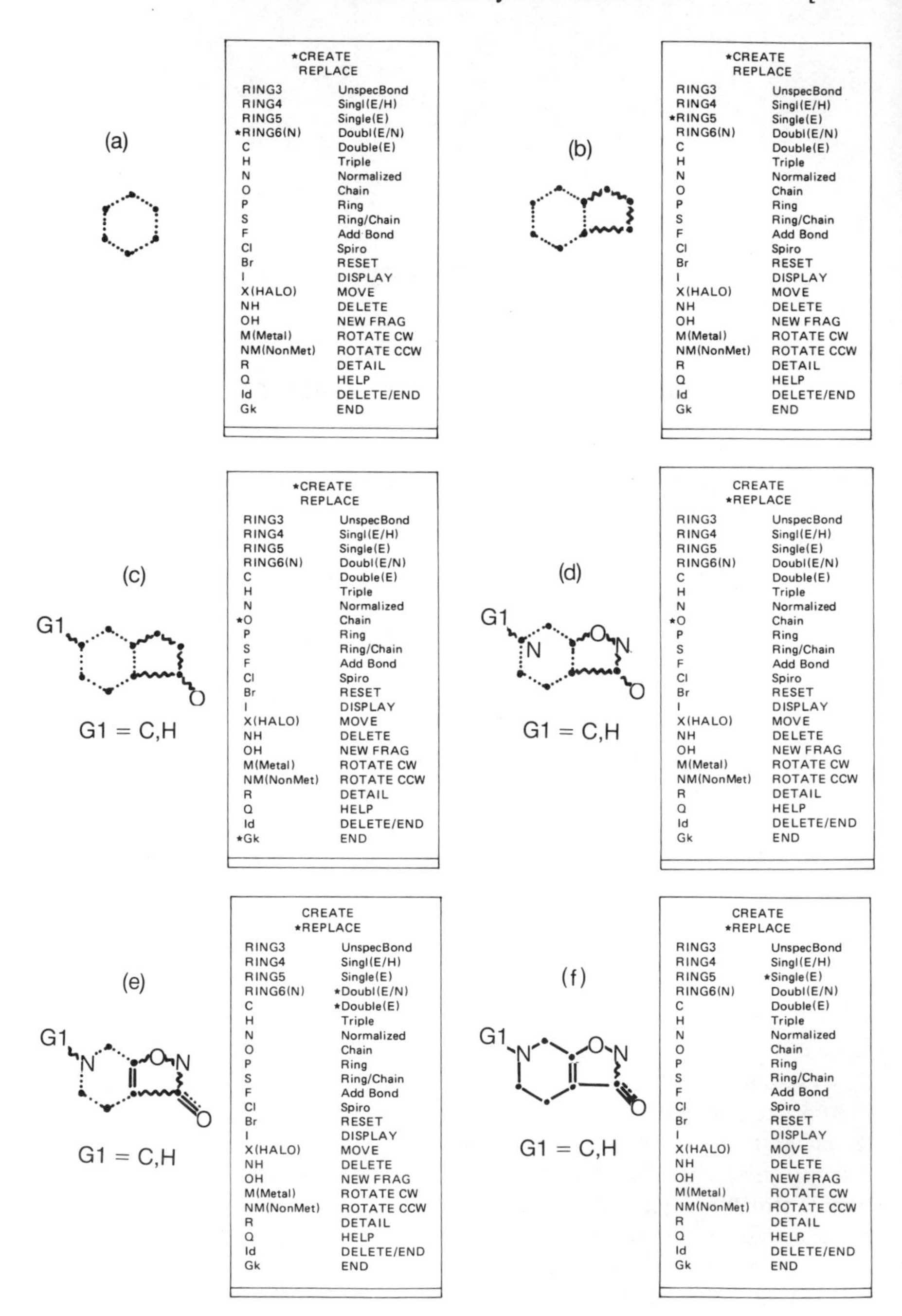

Fig. 7.4 – CAS ONLINE Graphic Structure Input (GSI). (CAS ONLINE Guide to Commands. Copyright by the American Chemical Society, reprinted by permission.)

so that it fuses with the ring already on the screen. The exocyclic oxygen is attached in Fig. 7.4(c), and the variable substituent *G1* is attached and defined as having the values C or H. Ring hetero-atoms are selected and positioned in Fig. 7.4(d), and normalized double bonds (Double (*E*/*N*)) and exact single bonds (Single (*E*)) are positioned in Figs. 7.4(e) and 7.4(f) respectively.

The 'G_k' group of CAS ONLINE, illustrated in this example, enables up to four variable atom groups to be defined in a query structure, each containing a maximum of four element symbols. In this respect, Télésystèmes-DARC is rather more flexible than CAS ONLINE [42]; for example, alternative bond types may be specified between selected atoms, although as stated above no distinction is made between chain and ring bonds.

The Graphic Substructure Query Module of COUSIN provides significantly greater flexibility for describing generic substructure queries by means of its 'R_k' notation [30]. Fig. 7.5 illustrates the construction sequence for a sample query. The ring skeleton and substituent groups are entered by means of a light-pen in the 'Draw Substructure' and 'Enter R Groups' steps, Figs. 7.5(a) and 7.5(b) respectively. The 'Draw Substructure' step enables predrawn ring and functional group templates to be displayed, atoms to be assigned and new bonds to be created, the latter being drawn on the screen following the motion of the light-pen. The 'Enter R Groups' step enables the searcher to indicate positions in the query where choices of atoms, groups or larger structural features are allowed. The 'R_k' groups are defined subsequently in Figs. 7.5(d) and 7.5(f) by supplying appropriate values and occurrence counts drawn in the blank regions of the screen, Figs. 7.5(c) and 7.5(e). The 'Bond Types' step enables the searcher to generalize or tighten constraints on individual bonds. In Fig. 7.5(g) the searcher has selected the appropriate bond types from the menu to indicate that the bond connecting the lower side chain to the indole ring must be acyclic (slash symbol, '/') and that the next bond must be in a ring (circle symbol, '○') and may have any order (dotted bond symbol, '. . .'). The 'Ring Counts' step allows the number of rings that must appear in retrieved compounds and their respective sizes to be specified, as in Fig. 7.5(h).

Of the remaining systems, MACCS provides the most sophisticated structure entry and image manipulation facilities. The MACCS structure input software was originally written for GT40 terminals and certain Tektronix terminals, although MDL have recently released a version of MACCS for use with the less expensive DEC VT640 terminals. DARC uses Tektronix or Tektronix-emulating graphics terminals and provides fast and efficient structure input but without some of the screen manipulation facilities of MACCS. CAS ONLINE graphic structure input is possible only on Hewlett-Packard HP2647A and HP2647F terminals, which must be programmed locally with a program tape supplied by CAS.

The observable trend among structure search systems is towards a more flexible interface with the user of the system. This is manifested in the interactive query

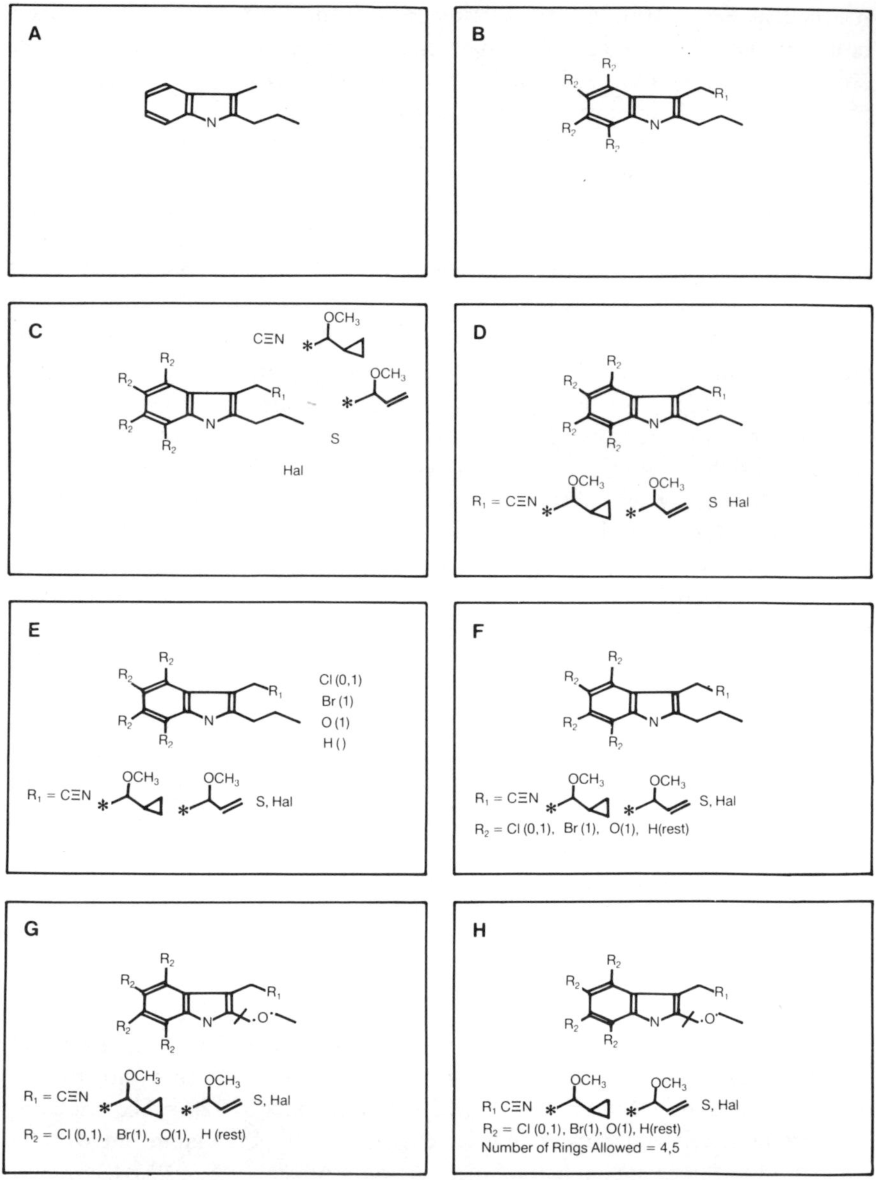

Fig. 7.5 – COUSIN graphic substructure query input. (Reprinted with permission from ref. [30]. Copyright 1982 American Chemical Society.)

modules described above, and in the provision of terminal software which enables manipulation of structural data and local reformatting of search output. In-house systems in particular have taken advantage of new hardware technology and database management techniques to achieve the integrated processing of proprietary and published data files, and include software modules for a range of applications in addition to compound registration, storage and retrieval. These applications include structure–activity correlation, structure elucidation, molecular modelling and computer-assisted synthesis design.

The diverse nature of these systems is a consequence of the history of their development. For example, the DARC code is a radically different form of structure representation to the connection table representations common to other systems. Even here, the notation-based CROSSBOW connection table differs significantly from the various forms of bond-explicit connection table described in Chapter 5. Despite these differences, and the differing approaches adopted for structure registration, search and display, these systems provide retrieval facilities which are broadly comparable, and a preference for any one system is likely to be made on considerations such as the ease of installation and availability of maintenance contracts, the integration of existing services and where necessary the conversion of company data files, and the nature of any applications software.

In the case of public-access systems, these too are broadly comparable in performance though not in design. The development paths of Télésystèmes-DARC and CAS ONLINE are now observed to be converging, each towards a software system capable of searching large files of registry data together with subsequent retrieval of associated bibliographic information from published databases. Here the choice of service will depend upon factors such as the pricing structure, the ease of use and familiarity with the system command language and query input modules, and the range of data files available for search.

REFERENCES

[1] Nübling, W., and Steidle, W., The Dokumentationsring der Chemisch-pharmazeutischen Industrie: Aims and Methods, *Angew. Chem. Internat. Edit.*, **9**, 596–598 (1970).

[2] Meyer, E., The IDC System for Chemical Documentation, *J. Chem. Doc.*, **9**, 109–113 (1969).

[3] Fugmann, R., The IDC System, in: *Chemical Information Systems*, Ash, J. E., and Hyde, E. (eds.), Ellis Horwood, Chichester (1975).

[4] Basel Information Center for Chemistry (Documentation Centre of Ciba-Geigy Ltd, F. Hoffmann–La Roche & Co., Ltd, and Sandoz Ltd), CH-4002 Basel, Switzerland.

[5] Buckley, J. S., Chemical Substructure Searching using Ringdoc/Ringcode, *Drug Information Journal*, **8**, 105–110 (1974).

[6] Oatfield, H., The ACRS System: Ringdoc as Used with a Computer, *J. Chem. Doc.*, **7**, 37–43 (1967).

[7] Deforeit, H., Caric, A., Combe, H., Leveque, S., Malka, A., and Valls, J., CORA: Semiautomatic Coding System. Application to the Coding of Markush Formulas, *J. Chem. Doc.*, **12**, 230–233 (1972).

[8] Bawden, D., and Devon, T. K., Ringdoc: The Database of Pharmaceutical Literature, *Database*, **3**, 29–39 (1980).

[9] Rössler, S., and Kolb, A., The GREMAS System, an Integral Part of the IDC System for Chemical Documentation, *J. Chem. Doc.*, **10**, 128–134 (1970).

[10] Graf, W., Kaindl, H. K., Kniess, H., Schmidt, B., and Warszawski, R., Substructure Retrieval by means of the BASIC Fragment Search Dictionary Based on the Chemical Abstracts Service Registry III System, *J. Chem. Inf. Comput. Sci.*, **19**, 51–55 (1977).

[11] Graf, W., Kaindl, H. K., Kniess, H., and Warszawski, R., The Third BASIC Fragment Search Dictionary, *J. Chem. Inf. Comput. Sci.*, **22**, 177–181 (1982).

[12] Farmer, N. A., and O'Hara, M. P., CAS ONLINE. A New Source of Substance Information from Chemical Abstracts Service, *Database*, **3**, 10–25 (1980).

[13] Dittmar, P. G., Farmer, N. A., Fisanick, W., Haines, R. C., and Mockus, J., The CAS ONLINE Search System. I. General System Design and Selection, Generation, and Use of Search Screens, *J. Chem. Inf. Comput. Sci.*, **23**, 93–102 (1983).

[14] Dubois, J. E., and Viellard, H., Système DARC. Théorie de Génération-Description, *Bull. Soc. Chim. France*, 900–919 (1968).

[15] Dubois, J. E., Anselmini, J. P., Chastrette, M., and Hennequin, F., Système DARC. Théorie de Population-Correlation, *Bull. Soc. Chim. France*, 2439–2448 (1969).

[16] Dubois, J. E., and Laurent, D., Système DARC. Théorie de Population-Correlation, *Bull. Soc. Chim. France*, 2449–2455 (1969).

[17] Dubois, J. E., and Viellard, H., Système DARC. Représentation des Compose Cycliques, *Bull. Soc. Chim. France*, 839–848 (1971).

[18] Dubois, J. E., French National Policy for Chemical Information and the DARC System as a Potential Tool of this Policy, *J. Chem. Doc.*, **13**, 8–13 (1973).

[19] Attias, R., DARC Substructure Search System: A New Approach to Chemical Information, *J. Chem. Inf. Comput. Sci.*, **23**, 102–108 (1983).

[20] Feldmann, R. J., and Heller, S. R., Application of Interactive Graphics: Nested Retrieval of Chemical Structures, *J. Chem. Doc.*, **12**, 48–53 (1972).

[21] Feldmann, R. J., Milne, G. W. A., Heller, S. R., Fein, A. E., Miller, J. A., and Koch, B., An Interactive Substructure Search System, *J. Chem. Inf. Comput. Sci.*, **17**, 157–163 (1977).

[22] Heller, S. R., Milne, G. W. A., and Feldmann, R. J., A Computer-based Chemical Information System, *Science*, **195**, 253–259 (1977).

[23] Milne, G. W. A., Heller, S. R., Fein, A. E., Frees, E. F., Marquart, R. G., McGill, J. A., Miller, J. A., and Spiers, D. S., The NIH-EPA Structure and Nomenclature Search System, *J. Chem. Inf. Comput. Sci.*, **18**, 181–186 (1978).

[24] Milne, G. W. A., and Heller, S. R., NIH-EPA Chemical Information System, *J. Chem. Inf. Comput. Sci.*, **20**, 204–211 (1980).

[25] Eakin, D. R., Hyde, E., and Palmer, G., The Use of Computers with Chemical Structural Information: ICI CROSSBOW System, *Pesticide Science*, **5**, 319–326 (1974).

[26] Eakin, D. R., The ICI CROSSBOW System, in: *Chemical Information Systems*, Ash, J. E., and Hyde, E. (eds.), Ellis Horwood, Chichester (1975).

[27] Townsley, E. E., and Warr, W. A., Chemical and Biological Data: An Integrated Online Approach, *ACS Symposium Series*, **84** (1978).

[28] Warr, W. A., ICI's Experiences with CROSSBOW, *Proc. 5th Int. Online Inf. Meeting*, Oxford, Learned Information (1981).

[29] Howe, W. J., and Hagadone, T. R., Progress toward an Online Chemical and Biological Information System at the Upjohn Company, *ACS Symposium Series*, **84** (1978).

[30] Howe, W. J., and Hagadone, T. R., Molecular Substructure Searching: Computer Graphics and Query Entry Methodology, *J. Chem. Inf. Comput. Sci.*, **22**, 8–15 (1982).

[31] Hagadone, T. R., and Howe, W. J., Molecular Substructure Searching: Minicomputer-Based Query Execution, *J. Chem. Inf. Comput. Sci.*, **22**, 182–186 (1982).

[32] Dill, J. D., Hounshell, W. D., Marson, S., Peacock, S., and Wipke, W. T., Search and Retrieval using an Automated Molecular Access System, paper given at 182nd National Meeting of the American Chemical Society, New York, August (1981).

[33] Polton, D. J., Installation and Operational Experiences with MACCS (Molecular Access System), *Online Review*, **6**, 235–242 (1982).

[34] Garfield, E., Revesz, G. S., Granito, C. E., Dorr, H. A., Calderon, M. M., and Warner, A., Index Chemicus Registry System: Pragmatic Approach to Substructure Chemical Retrieval, *J. Chem. Doc.*, **10**, 54–58 (1970).

[35] Ash, J. E., Connection Tables and their Role in a System, in: *Chemical Information Systems*, Ash, J. E., and Hyde, E. (eds.), Ellis Horwood, Chichester (1975).

[36] Dittmar, P. G., Stobaugh, R. E., and Watson, C. E., The Chemical Abstracts Service Chemical Registry System. I. General Design, *J. Chem. Inf. Comput. Sci.*, **16**, 111–121 (1976).

[37] Mockus, J., and Stobaugh, R. E., The CAS Chemical Registry System. VII. Tautomerism and Alternating Bonds, *J. Chem. Inf. Comput. Sci.*, **20**, 18–22 (1980).

[38] Elder, M., The Conversion from Wiswesser Line Notation to CIS Connection Tables, paper given at CNA(UK) Seminar on Interconversion of Structural Representations, Loughborough University, March (1982).

[39] Dittmar, P. G., Mockus, J., and Couvreur, K. M., An Algorithmic Computer Graphics Program for Generating Chemical Structure Diagrams, *J. Chem. Inf. Comput. Sci.*, **17**, 186–192 (1977).

[40] *CAS ONLINE Screen Dictionary for Substructure Search*, 2nd edn., Chemical Abstracts Service, Columbus, Ohio (1981).

[41] *Guide to CAS ONLINE Commands*, Chemical Abstracts Service, Columbus, Ohio (1981).

[42] Watson, D., Some Experiences with the DARC System, paper given at CNA(UK) Seminar on Chemical Structure Searching of the Published Literature, Daresbury, March (1980).

8

Reaction indexing

8.1 INTRODUCTION

The need for adequate means of retrieval for chemical reaction data has been apparent for many years now. The preface to the first edition of Weyl's classic work on organic chemistry [1] contained the statement that a scientist could hardly hope to be familiar with every one of the innumerable methods described therein, while, more recently, both Meyer [2] and Valls [3] have called attention to the importance of providing access to reaction information. As there are by now over six million compounds known, and as any one of them can be transformed by suitable reactions into a large number of others, it is clear that the volume of potential data is quite enormous, even if those classes of reactions are ignored that are known to exist, but for which there are currently no known members [4]; additionally, the number of reactions which have actually been reported in the literature is increasing rapidly [5]. Yet, despite the urgency of the need, there are currently few aids to help the chemist in his search for viable synthetic pathways: this lack may be in part responsible for the widespread recognition of the achievements of synthetic chemists such as Corey and Woodward, for the frequent use of terms such as 'elegant' in reviews of syntheses, and the calling of synthetic organic chemistry 'an art in the midst of a science' [6].

One of the main problems in devising a system for chemical reaction indexing is that, whereas a chemical molecule is a unique entity, and thus susceptible to listing in a canonical form, such as a systematic name or linear notation, a reaction has many characteristics, all of which may need to be stored for subsequent retrieval. Thus starting materials, products, reaction sites, catalysts, yields, mechanisms, experimental conditions, bond changes, and by-products may all need to be represented in a file for search purposes. Many of these features may be indexed via either free text or some form of encoding, and then retrieved using standard bibliographic means, while the reactant and product molecules may be handled using one of the standard means of representing chemical structures. Much more difficult, and the crux of the reaction indexing problem, is the provision of a machine-readable representation of the actual changes which have taken place in the course of the reaction. This chapter is primarily concerned with the characterization of these changes, i.e. of the transformation that has occurred in the course of a reaction, without regard to the actual substrate in which the transformation has taken place.

Three main approaches have been suggested for the characterization of transformations, these being the use of nomenclature, of mechanistic descriptors, and of the substructural changes engendered by the transformation. This chapter includes examples of all three indexing approaches and discusses the implementation of retrieval systems for reactions. It should be noted that the systems described here for reaction storage and retrieval are, in large part, experimental in character, and are thus in marked contrast to the long-established systems for structure storage and retrieval described in Chapters 5 to 7. Accordingly, this chapter concentrates upon the principles underlying the techniques that have been described to date, and which may form the basis for systems in the future, rather than describing currently available packages as has been done in earlier chapters.

8.2 A SYSTEMATIC NOMENCLATURE FOR ORGANIC TRANSFORMATIONS

As with the characterization of individual chemical compounds, the earliest forms of reaction index were based upon nomenclature, and to this day the most widely employed and most easily understood description is the use of a trivial name, usually that of the chemist or chemists who originally discovered the reaction. Terms such as the Fischer indole synthesis, the Claisen rearrangement, and the Wolff–Kishner reduction are common in the literature, and several compendia of named reactions are now available. Such nomenclature may prove useful in providing a concise description of complex reactions that are difficult to index by more systematic means, such as the Beckmann rearrangement, but, in general, the use of indexing terms which have no direct relationship with the reaction that they are supposed to describe leads to severe problems in retrieval. Thus structurally similar transformations may be separated which might be considered more fruitfully in conjunction, and there may also be disagreement as to the exact extent of the reactions that should be considered under a given heading [7]. However, the greatest deficiency in such a system is the lack of coverage since the great bulk of reaction types do not have a suitable name.

Some years ago, Bunnett [8] introduced a simple but systematic nomenclature for substitutions in which, for example, the transformation

$$C_6H_5{-}SiMe_3 + Br_2 \longrightarrow C_6H_5{-}Br$$

would be regarded as an example of a bromo-desilylation. When the International Union of Pure and Applied Chemistry (IUPAC) Commission on Physical Organic

Chemistry was formed in 1974, one of the first tasks it set itself was to develop this simple system to encompass as wide a range of types of transformation as possible. The system is designed to be used in written and spoken communication: for these purposes the names should be short and euphonious, should contain features distinctive to the ear, and should be adaptable to various languages. The resulting system of nomenclature has two main parts [9]. Firstly, the essential components of the name should be the names of the incoming and outgoing groups, together with a characteristic suffix to denote the type of transformation; the suffix is a vital component since names such as hydroxy-debromination, for the substitution of Br by OH, hydroxy-bromination, for the addition of OH and Br, and dehydroxy-debromination, for the elimination of OH and Br, do not discriminate sufficiently between the three, very different types of reaction. Secondly, three other types of information that may need to be included are the sites in the substrate at which there are changes in connectivity (although in the simplest cases, such as addition to a carbon–carbon double bond, these can be understood), varying degrees of specificity about the structure of the substrate and the incoming and outgoing groups, and information about the stereochemistry of the transformation.

8.2.1 Univalent addition, elimination and substitution transformations

The system may be illustrated by the following examples and the comments on them.

The equilibrium reaction

$$CH_2{=}CH_2 \rightleftharpoons CH_3{-}CH_2Br$$

would be described by the names hydro-bromo-addition and hydro-bromo-elimination for the forward and backward reactions respectively, with the addend or eliminand of lower priority according to the Cahn–Ingold–Prelog sequencing rules being named first. The hyphens could, of course, be omitted from everyday writing, and the name could be inverted for indexing purposes as in addition-hydro-bromo. This example has assumed that the addition is to a carbon–carbon multiple bond but in other cases it is necessary to specify the nature of the reacting sites in the substrate as in

$$CH_2{=}O \longrightarrow C_4H_9{-}CH_2{-}O{-}Li$$

for which specific and generic names are *O*-lithio-*C*-butyl-addition and *O*-metallo-*C*-alkyl-addition respectively. The substrate can also be included as an optional extension to the name as in

$$CH_3CH{=}CH_2 \longrightarrow CH_3CHBrCH_3$$

which would be called 1-hydro-2-bromo-addition to propane. Similarly, other information about the transformation may be included at the author's discretion as in

$$\begin{matrix} H & & COCH_3 \\ & C{=}C & \\ CH_3 & & Br \end{matrix} \longrightarrow CH_3{-}C{\equiv}C{-}COCH_3$$

which could be called *anti*-hydro-bromo-elimination. For addition to, or elimination from, non-adjacent sites it is necessary to include some designation of the atoms involved. Although there is no difficulty with a specified substrate as in

$$CH_3CH{=}CH{-}CH{=}CHCH_3 \longrightarrow CH_3CH(Br){-}CH{=}CH{-}CH(Br)CH_3$$

which would be called 2,5-dibromo-addition to hexa-2,4-diene, the designation of relative site positions is a potential source of confusion in generalized names, and it is hence suggested that a number followed by a solidus, i.e. the symbol '/', be used to signify a relative position. The last example hence becomes a member of the general class of 1/4/dibromo-additions.

Substitution transformations present a slight problem since Bunnett's nomenclature for these is well established by now, but is not entirely compatible with the new scheme. Accordingly, it is suggested that names for indexing purposes should be in the new form, but that names for use in speech or writing may follow the original suggestions so that, for example, the transformation

$$CH_3{-}Br \longrightarrow CH_3{-}OH$$

could be indexed under hydroxy-de-bromo-substitution, and referred to in normal communication as hydroxy-debromination. Moreover, as is common practice, one could omit the name of hydrogen as an entering or leaving group from the common names so that

$$C_6H_6 \longrightarrow C_6H_5{-}NO_2$$

and

$$C_6H_5{-}N_2^+ \longrightarrow C_6H_6$$

would be called nitration and dediazoniation in speech or writing, but nitro-de-hydro-substitution and hydro-de-diazonio-substitution in an index.

8.2.2 Other types of transformation

The examples given so far cover the first phase of the Commission's work in this area, dealing with univalent additions, eliminations, and substitutions [9]. As the work progressed, however, it became clear that the basic rules could be extended to cover a wider variety of transformations. For multivalent transformations, it is suggested that a multiplicity flag should be used as in

$$R{-}C{\equiv}N \longrightarrow R{-}C({=}O){-}NH_2$$

and

$$R{-}C{\equiv}N \longrightarrow R{-}C({=}O){-}OH$$

which would be called *NN*-dihydro-*C*-oxo-biaddition and hydroxy,oxo-de-nitrilo-tersubstitution respectively. These prefixes are not essential but they serve to draw attention to the nature of the transformation. The second of these two examples also shows that one can name a transformation as a substitution, even though the actual reaction may not be a simple displacement. The names are deliberately non-mechanistic but not anti-mechanistic. Thus, if it was known that a reaction involved the cleavage of some specific bond, the corresponding transformation would not be named in a way that implied that some other bond had been broken: for example, the first of the above two reactions would not be referred to as a carboxamido-de-cyano-substitution.

In some reactions, more than one of the reacting molecules may reasonably be considered to be the substrate, and in this case two transformations should be named as in

$$C_6H_5NH_2 + C_6H_5COCl \longrightarrow C_6H_5NH{-}COC_6H_5$$

which would be indexed under both *N*-benzoylation and anilino-dechlorination. However, there are cases involving two identical reactant molecules where the designation of one as a substrate and the other as a reagent is too artificial: to overcome this problem, it is suggested that transformations such as

$$2\ ArBr \longrightarrow Ar{-}Ar\ ,$$

$$2\ R_2C{=}O \longrightarrow R_2C(OH){-}C(OH)R_2\ ,$$

and

$$2\ EtOH \longrightarrow (EtO)_2CH_2$$

should be called debromo-coupling, 1/*O*-hydro-2-coupling, and *bis*-*O*-dehydro-methylene-coupling respectively.

Other classes of transformation which are currently being studied include attachments and detachments, exemplified by the transformations

$$Me_3N \longrightarrow Me_3N^+{-}O^-$$

and

$$CH_2N_2 \longrightarrow CH_2$$

which would be called *N*-oxygen-attachment and dinitrogen-detachment respectively, insertions and extrusions, ring openings and closures, and rearrangement transformations. The Commission is still working on these problems and is actively seeking comments on the work that has been carried out to date. Once the work has been completed, it is hoped that the resulting system of names will prove to be a useful and practical tool for chemical reaction indexing.

8.3 A SYSTEMATIC NOMENCLATURE FOR ORGANIC MECHANISMS

Apart from the work on naming transformations, the IUPAC Physical Organic Commission has also been studying schemes for the systematic naming of reaction mechanisms [10]. Several such schemes have been reported in the literature [11–13] but the IUPAC study represents a much more detailed and systematic approach to the problem than hereto.

The proposed system has two parts, these being a simple but logical scheme which is intended to replace the Ingold S_N1, S_N2 scheme and its various extensions for common simple systems, and a detailed scheme that is intended to provide a notation for mechanisms of any degree of complexity: it is this latter work that is described here. A mechanism is taken to be a detailed description of a particular reactant to product path, together with information pertaining to intermediates, transition states, stereochemistry, the rate-limiting step, electronic excitation and transfer, and the presence of any loose or intimate electron ion pairs. All Markush structures undergoing the same process have the same mechanism, and are accordingly assigned the same symbolism.

The requirements of a systematic notation for reaction mechanisms are manifold: among the more important points are that it should be capable of describing the mechanistic behaviour as evidenced by product, kinetic and labelling studies, allow a ready correlation with conventional reaction diagrams, and thus permit the description of bond changes and atom and electron shifts. Additionally, it should allow for facile input and output via computer peripherals, and be usable for literature searching so that KWIC-type indexes, for example, can be readily generated to allow searches for textual fragments within the notational string. Finally, and most importantly, it should be capable of extension or abbreviation as mechanistic knowledge increases, and be acceptable to practising synthetic chemists.

If a representation of a mechanism is to be successful, it must be organized so as to reflect the way in which the basic mechanistic information is structured. Thus, since a mechanism may generally be subdivided into elementary reactions, and each of them may be further subdivided into primitive changes, so the representation for storage and retrieval should have a similar hierarchical format. In the notation, the name of the substrate, or, more precisely, the minimal structure that contains the reaction centre, is followed by a list of elementary reactions, each of which consists of a string of primitive change representations operating on specific atoms. A prefix field is included with each elementary reaction to include other data such as stereochemistry, indication of the rate-limiting step, and the sequence of elementary reactions. Elementary reactions are separated by the '+' symbol.

In order to implement such a notation, definitions are needed for each of the symbols that are to be used; the present state of the IUPAC work is that the general outline of the notation is provisionally agreed, but that discussions are proceeding gradually about the detailed symbols and their applicability.

A few typical examples are shown in Fig. 8.1. The primitive change identifiers are one or more letters, beginning with an upper-case one, with preference being given to symbols with some mnemonic value and followed by details of the atoms on which they operate. The sequence in which the atomic symbols are written in the parentheses is a useful indicator of the direction of electron movement. There are also requirements for symbols for primitive changes which do not involve bond breakages. The list is of course open-ended since new types of primitive change may await discovery.

Primitive changes mainly operate on two atoms

e.g. A(N, B) $NH_3 + BF_3 \longrightarrow H_3NBF_3$

Ac(C, C) $CH_3^{\cdot} + CH_3^{\cdot} \longrightarrow C_2H_5$

A = association C = colligation, c.f. also
D = dissociation, p = π-bond involved, r = odd no. of electrons.

but some of them operate on more

e.g. U (C, 2/C, H) CH_3–CH_3 + H^+ ⟶ CH_3⋯CH_3 + H (bridged)

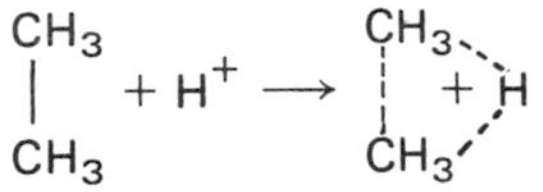

I (R, H, C) CH_2 + R–H ⟶ R–CH_2–H

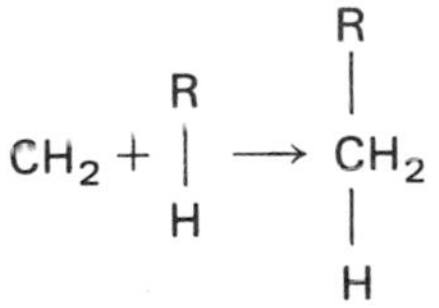

AAcy (1/C, 1/C, 4/C, 2/C) diene (1–4) + ethylene (1–2) ⟶ cyclohexene

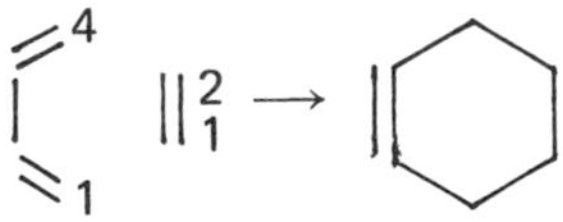

T (es, C) $CH_2{=}CH_2 + e^-_{(solv)} \longrightarrow [CH_2{-}CH_2]^{\overline{\cdot}}$

U = union, I = insertion T = electron Transfer with the electron treated as a reagent

Fig. 8.1 – Examples of the symbols used for primitive changes.

Suggestions have been made to define how to put together groups of primitive changes in a unique manner, to identify and number the reactant centre, to specify stereochemistry, etc. An example of the resulting notational details is given in Fig. 8.2, and the IUPAC Commission would welcome detailed comments on these proposals. The pinacol rearrangement is presumed to be a three-step mechanism, although two-step or four-step processes are, of course, also conceivable. The first step is the association of a particular oxygen atom situated on 1/C, that is carbon atom number one in a relative notation, acting as the nucleophile, with a hydrogen species as the electrophile, that is H^+ of unspecified origin. The second step is the rearrangement proper, involving four bond changes specified as associations and dissociations between the indicated atoms, with the electron movement being from left to right. It should be noted that the pi bond formation between oxygen and carbon is represented by Ap, the rate-limiting step by #, and the stereochemical requirements by the term 'trans'. It is also possible to produce the very simple minimal structure shown, since three of the CH_3 groups are not involved in the bonding changes, even though the compound of the same name may well not undergo the pinacol rearrangement.

Pinacol rearrangement

$HOC(CH_3)_2–C(CH_3)_2–OH + H^+$

Step 1 $\rightleftharpoons HOC(CH_3)_2–C(CH_3)_2–\overset{+}{O}H_2$

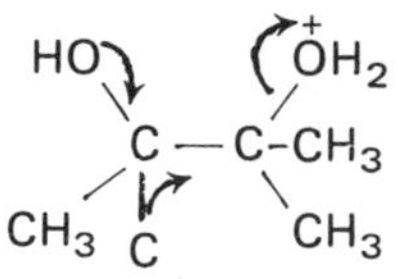

HO, $\overset{+}{O}H_2$, C — C-CH_3, CH_3, C, CH_3

Step 2

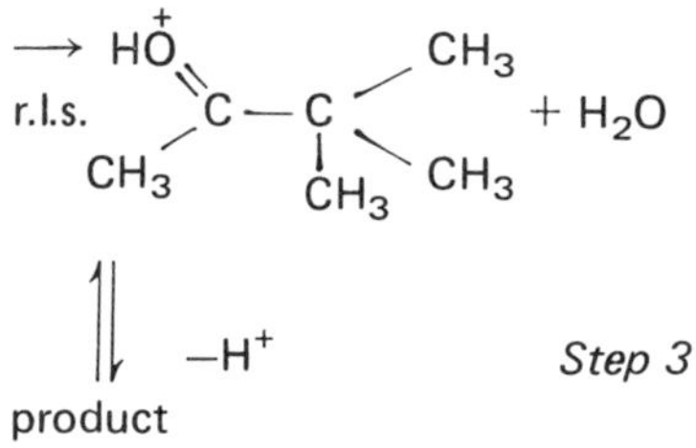

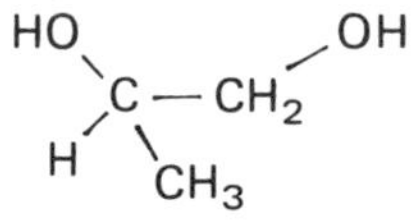

→ r.l.s. $H\overset{+}{O}$=C — C, CH_3, CH_3, CH_3, CH_3 + H_2O

⇅ $-H^+$ *Step 3*

product

HO, OH, C — CH_2, H, CH_3 — Minimal structure

Minimal structure name: 1,2-dihydroxy propane;

Step 1 notation: A(O[1/C],H) +

Step 2 notation: #, trans: Ap(O[2/C],2/C)D(3/C,2/C)
A(3/C, 1/C)D(O[1/C],1/C) +

Step 3 notation: D(O[2/C],H)

Fig. 8.2 – Example of a notation for the pinacol rearrangement.

The advantages of the proposed notation are several-fold. Firstly, there is a close relationship with the conventional reaction diagram with mechanistic data included. Secondly, indexing is possible by a very wide range of attributes including the substrate, the main groups of primitive changes and also the individual primitive changes, the subgroups of primitive changes using locant numbers with or without the corresponding atomic symbols, or the subgroups with related sets of locants; also, the inclusion of extra steps, as in steps 1 and 3 in Fig. 8.2, or of prefixes, does not invalidate the search, but may provide a means of locating an interesting subset of mechanisms in a file. Finally, it should be possible to locate not only all reactions with a given mechanism, such as the pinacol rearrangement in the example, but also closely related ones by widening the search: thus Wagner–Meerwein processes, Hofmann degradations and other 1,2 shifts, concerted or not with the loss of the leaving group might also be retrieved. Finally, the association and dissociation terms are closely related to the structural differences between the reacting molecules, and it will be shown below that such differences may be readily identified by automatic means: there is thus the possibility of automatic conversion between the two types of reaction descriptor.

8.4 STRUCTURE-BASED INDEXING OF REACTION INFORMATION

Yet another approach to the characterization of reactions is based on the structural differences between the reactant and product molecules, and involves a classification of reactions on the basis of the overall structural change taking place. Structure-based indexing methods are limited in that the initial and final states of the reacting molecules may not adequately describe the exact nature of the changes which are taking place. Thus an ester may be hydrolysed to the corresponding acid by the rupture of either an alkyl-to-oxygen or an acyl-to-oxygen bond: however, the overall transformation is the same in both cases. That said, this information is often sufficient for retrieval purposes in synthetic chemistry; additionally, the changes may be determined by purely automatic means from the structures of the reacting compounds.

8.4.1 Automatic methods for the detection of reaction sites

Structure-based indexing requires the development of effective and efficient algorithms for the identification of the reaction sites, these being the substructures in the reacting molecules which have been directly involved in the reaction. The identification is done by means of an automatic comparison of the machine-readable structure representations of the reactant and product molecules so as to identify the differences in structure: these differences may then be used as the basis for subsequent retrieval. This relatively simple approach has formed the basis for a research project in the University of Sheffield which has tested a range of methods for identifying the reaction sites [14–18].

The methods are based on the use of reacting molecules encoded in either Wiswesser Line Notation (WLN) or connection tables, these being the two main types of structure representation encountered in computer-based chemical information systems. The WLN approach [14] provides a cheap means of generating simple printed indexes of reactions and is based on differences in the WLN symbol strings of the reactant and product molecules. An evaluation of such a printed index showed that it offered a retrieval capability which was at least comparable with that obtainable from the commercially available Derwent Chemical Reactions Documentation Service [15].

A superior approach is based on connection tables [16–18] and this method is described here in some detail. But first, it is worth considering the most sophisticated form of analysis which a computer could reasonably be expected to carry out using structural information alone, and without the benefit of mechanistic knowledge. This analysis is the identification of the maximal common substructure (MCS) of the reactant and product molecules, and is simply that portion of the compounds common to each other: if the MCS can be identified, the reaction sites are those parts of the reacting molecules which are not included in the MCS. Although simple in concept, MCS identification is very time-consuming, involving as it does an amount of computation proportional to the product of the factorials of the numbers of atoms in the reacting compounds [18]. Such algorithms are hence quite unsuitable if thousands of reactions need to be processed, as would be required for the creation of a large reactions data base.

Since MCS detection is so time-consuming, a very simple, but approximate, structure-matching algorithm has been developed which finds a large common substructure, although not usually the maximal one [16]. This restriction is more than compensated for by a drastic increase in computational efficiency: specifically, the common substructures are identified about two orders of magnitude faster than for the detection of the MCS. Once the common substructures have been found, they may be eliminated and the remaining parts of the reactants and products taken to be the reaction sites: generally, the algorithm finds the MCS together with a few attached atoms and bonds. A similar procedure has recently been described by Jochum *et al.* [19].

The matching procedure [16] is derived from the Morgan algorithm which was developed at Chemical Abstracts Service as a means of obtaining canonical connection tables for storage in the Registry system. This algorithm is based upon the number of atoms immediately adjacent to each of the atoms in a molecule, the first-order connectivity. The second-order connectivities for an atom may be obtained by summing the first-order connectivities of the adjacent atoms, and similarly for the third, fourth ... order values. Writing $c_i(n)$ for the nth order connectivity of the ith atom in a molecule, these higher order connectivities may be obtained from

$$c_i(n) = \sum_j c_j(n-1) \quad ; \quad j \text{ adjacent to } i,\ n > 1 \ .$$

Owing to the manner in which the connectivities are calculated, the nth order value for an atom is a numerical identifier which characterizes a circular substructure of radius $n-1$ bonds. These identifiers may be made more discriminating by including, in the first order values, information about the elemental type of the central atom, and of the types of atom and bond immediately adjacent to it. Comparison of the nth order values for two atoms provides a rapid and simple test for the isomorphism, i.e. structural equivalence, of the two corresponding circular substructures of radius $n-1$ bonds. If the two values are not the same, then the corresponding circular substructures cannot possibly be isomorphic; conversely, if a pair of atoms share equal sets of values $c_i(m)$ for $1 \leqslant m \leqslant n$, the two substructures may be isomorphic. Whereas the Morgan algorithm was originally developed for use with single molecules, reactions involve groups of molecules and thus extended connectivity offers a rapid means of investigating intermolecular substructural equivalences by comparing the sets of values $c_i(n)$ for one side of the reaction equation with the corresponding sets for the other. It is known *a priori* that large substructural similarities are likely to exist between the reactants and products in a change, and the assumption is made that the detection of equal sets of property values denotes the presence of an isomorphism. The calculation of the property values is very much faster than the time-consuming, atom-by-atom searching which is required to establish absolutely the presence of an isomorphism, and the assumption hence permits the processing of very large files of reactions at an acceptable computational cost.

The operation of the procedure is illustrated by the functional group interconversion reaction shown in Fig. 8.3. The highest equal property values are

(a) $CH_3-C(=O)-CH_2-CH(CH_3)-CH(CH_2CN)-C(=O)-NH_2 \longrightarrow CH_3-C(=O)-CH_2-CH(CH_3)-CH(CH_2CN)-CN$

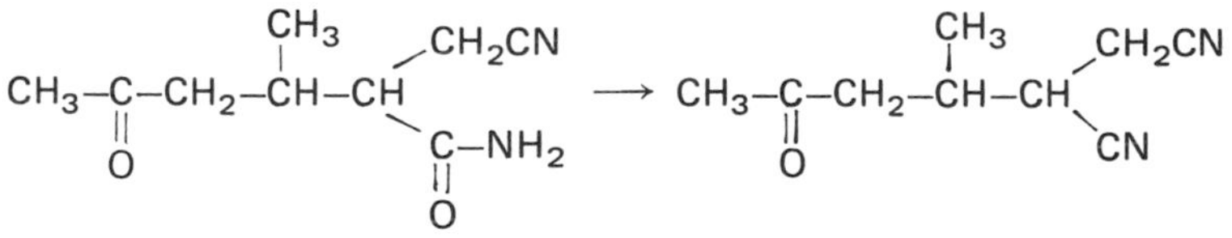

(b) $\diagup CH_2CN,\ \diagdown C(=O)-NH_2 \longrightarrow \diagup CH_2CN,\ \diagdown CN$

(c) $-C(=O)-NH_2 \longrightarrow -CN$

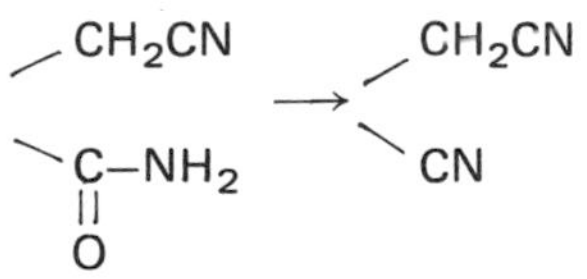

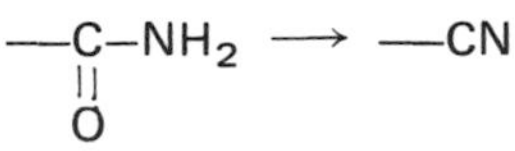

Fig. 8.3 – Automatic detection of reaction sites by an approximate structure-matching algorithm.

obtained with the fifth order values for the terminal methyl atoms in the reactant and product molecules of Fig. 8.3(a), and all of the atoms within four bonds of these atoms may be eliminated from consideration for inclusion in the reaction sites. The remaining substructures are shown in Fig. 8.3(b) and, of the atoms there, the greatest similarity is that between the nitrogen atoms in the cyanide grouping which have equal fourth order values. The three-bond radius circular substructures are eliminated to yield the substructures of Fig. 8.3(c). No further equivalences can be identified and the procedure terminates with these reaction sites as the result of the analysis. The use of the algorithm has thus resulted in a marked localization of the reaction site. A detailed failure analysis reveals some limitations in the procedure, but an intuitively reasonable reaction site is obtained in at least 93 per cent of the reactions studied in the course of the project [16, 17].

8.4.2 A matrix description of chemical reactions

If a representation of the reaction sites can be generated, a useful addition to any indexing system would be the provision of a unique and succinct notation to characterize the sites. This problem has been studied by Brandt and his coworkers [20] who have developed a hierarchical classification of reactions. The primary basis of the classification is the set of bonds that have been broken or created in the course of the reactions. For example, all reactions of the type

$$\text{A–B} + \text{C–D} + \text{E–F} \longrightarrow \text{F–A} + \text{B–C} + \text{D–E}$$

such as the Cope rearrangement, Diels–Alder reaction or pinacol rearrangement belong to the same category. This is, of course, a very general classification and more specific descriptors are derived in the second level of the classification which is based upon those bonds and free electrons in the reaction sites that remain unchanged during the course of the reaction. Thus all Cope rearrangements belong to one class

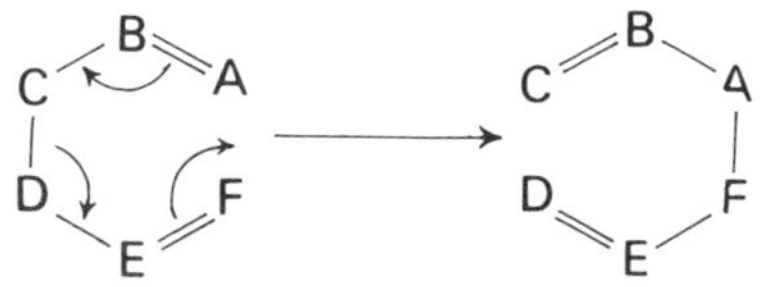

(8.1)

whereas all Diels–Alder reactions belong to another

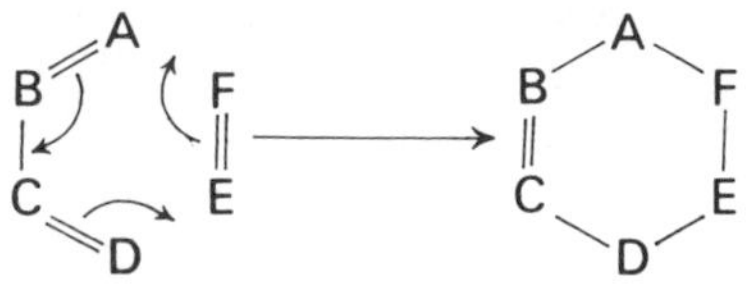

The third level involves a description of the atoms A, B ... F: this may be based just upon the atomic symbols or may involve taking the neighbourhood spheres into account. The mathematical model of constitutional chemistry first described by Dugundji and Ugi [21] forms a useful tool for the development of such classification schemes. In this model, a chemical reaction can be expressed by a master equation

$$B + R = E$$

where B and E are the bond–electron matrices of the ensembles of molecules (EM) that take part in the reaction. An EM may consist of one or more molecules. The matrix R is the reaction matrix which describes the rearrangement of the valence electrons, i.e. the pattern of electron shifts which take place in the course of the reaction. The off-diagonal entries, r_{ij}, of R indicate the bonds that have been created or broken, while the diagonal entries, r_{ii}, show the changes in the free electrons on the atoms a_i. Since the rearrangement of bonds and free electrons does not involve all of the atoms a_i, the matrix R will contain many columns and rows all of whose entries are zero-valued, and R is thus, in general, not completely suited to the purposes of reaction documentation and classification. A more suitable matrix is the irreducible R-matrix, R^I, which is obtained by eliminating all of the zero-valued rows and columns and retaining only those which contain at least some non-zero-valued entries. If the resulting matrix, R^I, contains n rows and columns, there are at most $n!$ distinct possible numberings of the elements. There are many ways of generating a canonical R-matrix, R^C, for use in a reactions retrieval system: the canonicalization rule chosen reflects some aspects of a chemist's visualization of valence electron shifts. In most reactions, bonds or electron pairs appear, at least in a formal sense, to be shifted along a path or along the circumference of a ring. Thus if the atoms in the reaction core are arranged in the sequence of the electron shifting path, the corresponding matrices that describe such electron shifts show alternating signs along the first and the outermost off-diagonal. Thus the canonical R-matrix for the reaction 8.1 above is

$$\begin{matrix} 0 & -1 & 0 & 0 & 0 & 1 \\ -1 & 0 & 1 & 0 & 0 & 0 \\ 0 & 1 & 0 & -1 & 0 & 0 \\ 0 & 0 & -1 & 0 & 1 & 0 \\ 0 & 0 & 0 & 1 & 0 & -1 \\ 1 & 0 & 0 & 0 & -1 & 0 \end{matrix}$$

In cases where more than one permutation can result in the same initial set of off-diagonal entries, the other diagonals of R^I will need to be considered in turn until a decision is reached. With these sequence rules, a very efficient algorithm can be formulated for the computation of the canonical numbering, which makes use of the algebraic properties of R-matrices [22]. This formalism allows one to write a new description of the first classification rule as 'All reactions with the same canonical R-matrix R^C belong to the same R-category'.

The second level in the classification is based upon the unchanged bonds in the reaction centre, and these may be described by the intact bond-electron-matrix, B^I. The B^I of the Diels–Alder rearrangement, for example, is

$$\begin{matrix} 0 & 1 & 0 & 0 & 0 & 1 \\ 1 & 0 & 1 & 0 & 0 & 0 \\ 0 & 1 & 0 & 1 & 0 & 0 \\ 0 & 0 & 1 & 0 & 0 & 0 \\ 0 & 0 & 0 & 0 & 0 & 1 \\ 1 & 0 & 0 & 0 & 1 & 0 \end{matrix}$$

In terms of the Dugundji–Ugi model, the second level of the classification may be formally expressed as 'All reactions with the same R^C and the same B^C belong to the same RB-category'. In many cases, the canonical numbering may also be obtained manually by simply looking up the longest alternating path (chain or ring).

Various extensions or improvements to the basic approach are possible. Thus, stereochemical features might be incorporated by appending parity vectors or other stereochemical descriptors based on the theory of chemical groups, while it might also be possible to include some representation of the mechanism by which the transformation has occurred. Finally, it may be noted that although the R-matrix contains the complete information concerning the transformation of one chemical structure into another, it is not very suitable for describing a reaction in a printed index or textbook. A linear notation can, however, be readily derived from the canonicalization rules by inserting separator characters between the different substrings and by introducing abbreviations to improve the legibility of the notation: the resulting string is much more suited to applications involving printed media.

8.5 STRUCTURE-BASED RETRIEVAL OF REACTION INFORMATION

The previous sections have described methods for the indexing of chemical reaction information, but indexing, whether manual or automatic, is only carried out to facilitate the subsequent retrieval of reactions from a database. This section considers some of the work currently in hand to develop search systems for reactions, and discusses problems that need to be overcome before large databases of reactions become generally available for search.

An important feature of a comprehensive retrieval package for reactions, and the feature that, perhaps, most clearly distinguishes such systems from those designed for the retrieval of molecules, is the need to provide modes of access for queries that involve the specification of both changed and unchanged substructural features. The reaction sites for a reaction are not sufficient in themselves for the complete representation of a change because some searches (for example a search for the reduction of a ring carbonyl during which an acyclic ketoester remains unchanged) cannot be carried out in the absence of information concerning the exact structural environment of the reaction sites. These types of query suggest that a retrieval system for reactions should be based upon four sets of chemical fragment screens, these being for the reactant and product reaction sites, and for the corresponding unchanged portions of the reacting compounds. Such a screening system would also be capable of dealing with less complex types of query, including the synthesis of some particular product molecule or substructure, the reactions of a given set of reactants, and substructural transformations in which the environment is not specified. Molecular retrieval systems, conversely, are based upon only a single set of fragment screens for each of the compounds in a file.

8.5.1 An experimental retrieval system for reactions

An evaluation has been reported of a retrieval system that is based upon the approximate structure matching algorithm described in 8.4.1 [17]. Once the reaction sites in a set of reacting molecules have been identified, the algorithm outputs the input connection tables, together with a series of flags denoting whether a particular atom is, or is not, included in the reaction centres. With these annotated connection tables, it is possible automatically to generate the aforementioned set of four fragment screens, and these are used for the characterization of the reactions in the file; in addition, a variety of molecular formula and WLN descriptors [14] are generated so as to provide a wide range of alternative modes of access. The evaluation involved the creation of a search file of over four thousand indexed reactions taken from the Institute for Scientific Information (ISI) product *Current Abstracts of Chemistry*, and a collection of over one hundred reaction queries culled from the synthesis design and reaction indexing literature, and from the research information departments of two United Kingdom pharmaceutical firms. This set of queries gave a mean screenout of over 99 per cent while the precision for those queries that retrieved some material was about 65 per cent, even without any facilities for atom-by-atom searching after the bit string search. These results would seem to imply that although the indexing method is only approximate in nature, it is sufficiently precise to allow searches to be carried out that are comparable in discrimination to those obtainable from conventional structure retrieval systems.

8.5.2 REACCS – a program package for the retrieval of reactions

Such ideas have recently been taken considerably further in the REACCS package which is marketed by Molecular Design Limited and which is based upon the use of structure-based methods of reaction indexing, of hierarchical data management techniques, and of the sophisticated structure input and display features characteristic of modern chemical graphics systems.

Reactions are entered into the database either from a disc file or via a computer graphics terminal; in the latter case, the reacting molecules are sketched using a light pen, while textual and numeric data are input from the keyboard. The reaction centres are identified automatically, and the resultant analysis presented to the chemist running the program to allow manual checking: this is especially important when non-stoichiometric reactions are processed by an automatic indexing procedure. A differentiation is made between the reacting and non-reacting atoms in the molecules, thus permitting queries involving both reaction sites and unchanged features. In fact, the reaction centres are highlighted when sets of molecules are displayed on the screen of the terminal. In addition to the structural features of a reaction, provision is made for much accessory reaction data in numerical or textual form. Thus it is possible to specify literature references, the type, volume and physical properties of the solvent, the pressure and temperature requirements of the reaction, etc. All of these data elements are searchable, and it is thus possible to carry out searches involving both structural and non-structural search parameters. Queries are input to the system in much the same way as a new reaction is incorporated, with the retrieved items being displayed at the terminal, or upon a hard copy output device, in real time.

It is likely that systems such as this will become increasingly available within the near future so that access to the reaction literature will become as easy as it already is to information on chemical structures.

8.6 PROBLEMS IN THE PROVISION OF REACTION INFORMATION

A problem with the use of reaction information is that there are currently no large public databases of reactions available in a machine-readable form that are suitable for processing and searching by systems such as REACCS. The Chemical Reaction Documentation Service produced by Derwent Publications Ltd involves the reacting molecules being encoded only with a manually assigned fragmentation code, while reaction site information is restricted to a note of the bonds that have been altered in the course of the reaction. Although the system offers an acceptable level of retrieval effectiveness [15], the lack of a whole structure representation for the reacting molecules means that the techniques described above cannot be employed. An alternative source of reaction information is the ISI publication *Current Chemical Reactions.* This contains novel reactions and syntheses taken from *Current Abstracts of Chemistry*, for which a magnetic tape is available containing the WLNs of all of the novel compounds in the printed publication. Unfortunately, the tape does not contain sufficient information to enable an automatic linking of the reactant and product molecules in each of the reactions that are reported, although this information could be obtained by manual inspection of the hard copy.

Even if one of these database producers were to undertake the marketing of a machine-readable file of reacting molecules, one further problem would need to be overcome. This factor, which has been alluded to in section 8.1, is the multi-faceted nature of reaction information which makes it difficult to obtain a single characterization for a reaction; accordingly it may prove difficult to find a registration procedure for determining whether a reaction is novel and should be included in the database. For example, an increase of 10 per cent in the yield of a reaction owing to some particular combination of pressure and temperature might well be of crucial importance in commercial circumstances, but the inclusion of large numbers of combinations of reaction conditions is not likely to be an option that commends itself to database producers. Again, a well-known transformation, but one that has been carried out in a novel substructural environment, might, or might not, be considered for inclusion. The problems are less obvious for an internal file of reactions since the particular areas of chemistry that are of interest are likely to be sufficiently limited to allow the inclusion of most, if not all, of the appropriate synthetic methods from laboratory notebooks or the printed literature.

REFERENCES

[1] Weyl, T. H., *Die Methoden der Organischen Chemie*, Leipzig, Georg Thieme (1901–1911).

[2] Meyer, E., Information Science in Relation to the Chemist's Needs, in Ash, J. E., and Hyde, E. (eds.), *Chemical Information Systems*, Chichester, Ellis Horwood (1975) .

[3] Valls, J., Reaction Documentation, in Wipke, W. T., Heller, S. R., Feldmann, R. J., and Hyde, E. (eds.), *Computer Representation and Manipulation of Chemical Information*, New York, Wiley (1973).

[4] Hendrickson, J. B., Systematic Synthesis Design. IV. Numerical Codification of Construction Reactions, *J. Am. Chem. Soc.*, **97**, 5784–5800 (1975).

[5] Garfield, E., Revesz, G. S., and Batzig, J. H., The Synthetic Chemical Literature from 1960 to 1969, *Nature*, **262**, 307–309 (1973).

[6] Hendrickson, J. B., Systematic Synthesis Design, *Top. Curr. Chem.*, **62**, 49–172 (1976).

[7] Fugmann, R., Kusemann, G., and Winter, J. K., The Supply of Information on Chemical Reactions in the IDC system, *Inform. Proc. Manag.*, **15**, 303–323 (1979).

[8] Bunnett, J. F., Systematic Names for Substitution Reactions, *Chem. Eng. News*, **32**, 4019 (1954).

[9] Bunnett, J. F., Nomenclature for Straightforward Transformations (Provisional), *Pure Appl. Chem.*, **53**, 305–321 (1981).

[10] Littler, J. S., An Approach to the Linear Representation of Reaction Mechanisms, *J. Org. Chem.*, **44**, 4657–4667 (1979).

[11] Mathieu, J., Allais, A., and Valls, J., Nucleofuger und Elektrofuger Austritt, *Angew. Chem.*, **72**, 71–74 (1960).

[12] Guthrie, R. D., A suggestion for the Revision of Mechanistic Descriptors, *J. Org. Chem.*, **40**, 402–407 (1975).

[13] Satchell, D. P. N., The Classification of Chemical Reactions, *Naturwissenschaften*, **64**, 113–121 (1977).

[14] Lynch, M. F., and Willett, P., The Production of Machine-readable Descriptions of Chemical Reactions using Wiswesser Line Notations, *J. Chem. Inf. Comput. Sci.*, **18**, 149–154 (1978).

[15] Bawden, D., Devon, T. K., Jackson, F. T., Wood, S. I., Lynch, M. F., and Willett, P., A Qualitative Comparison of Wiswesser Line Notation Descriptors of Reactions and the Derwent Chemical Reaction Documentation Service, *J. Chem. Inf. Comput. Sci.*, **19**, 90–93 (1979).

[16] Lynch, M. F., and Willett, P., The Automatic Detection of Chemical Reaction Sites, *J. Chem. Inf. Comput. Sci.*, **18**, 154–159 (1978).

[17] Willett, P., The Evaluation of an Automatically Indexed, Machine-readable Chemical Reactions File, *J. Chem. Inf. Comput. Sci.*, **20**, 93–96 (1980).

[18] McGregor, J. J., and Willett, P. Use of a Maximal Common Subgraph Algorithm in the Automatic Identification of the Ostensible Bond Changes Occurring in Chemical Reactions, *J. Chem. Inf. Comput. Sci.*, **21**, 137–140 (1981).

[19] Jochum, C., Gasteiger, J., and Ugi, I., The Principle of Minimum Chemical Distance (PMCD), *Angew. Chem. Int.*, **19**, 495–505 (1980).

[20] Brandt, J., Bauer, J., Frank, R. M., and von Scholley, A., A Classification of Reactions by Electron Shift Parameters. Application of the Dugundji–Ugi model, *Chem. Scripta*, **18**, 53–60 (1981).

[21] Dugundji, J., and Ugi, I., Algebraic Model of Constitutional Chemistry as a Basis for Chemical Computer Programs, *Top. Curr. Chem.*, **39**, 19–64 (1973).

[22] Brandt, J., and von Scholley, A., An Efficient Algorithm for the Computation of the Canonical Numbering of Reaction Matrices, *Comp. Chem.*, **7**, 51–59 (1983).

9

Techniques of structure manipulation

Previous chapters have discussed the classical, and predominantly manual, means that chemists use to acquire information, and the sophisticated computer systems that now permit the interactive searching of very large files of chemical structures. The ease with which chemical structures can be stored and manipulated in machine-readable files has led to the development of computer systems which *manipulate* structure information for purposes other than merely substance retrieval [1]. Section 9.1 of this chapter discusses one such manipulative technique, this being the prediction and correlation of physical, chemical and pharmacological properties with molecular structure. These methods are based in large part upon the use of computer graphics and of pattern recognition and multivariate statistical techniques, but there is now interest in the use of methodologies from the new and evolving discipline of *Artificial Intelligence, AI* [2]. As this name suggests, problem-solving programs attempt to find solutions to problems which are generally considered to require some degree of intelligence and specialist knowledge. *Expert systems* [3] are perhaps the most familiar of these programs, and systems for medical diagnosis [4] and the elucidation of molecular structure from analytical and spectroscopic data [5, 6] have already demonstrated levels of performance comparable to that of a human expert in their respective subject domains. A detailed description of DENDRAL, the most prominent example of a system for structure elucidation, has recently been published elsewhere [5]. Section 9.2 describes approaches to the application of computers in synthetic organic chemistry for designing chemical syntheses, and illustrates them with a number of synthesis design programs [7]. Both structure–activity correlation and synthesis design are problems of considerable topical interest in the pharmaceutical industry owing to the enormous costs involved in the development and testing of drug compounds.

9.1 STRUCTURE–ACTIVITY RELATIONSHIPS

9.1.1 Introduction

The identification of structure–activity relationships (SARs), involves the correlation of property data, such as physical, chemical or biological properties, with structural parameters characterizing the molecules in a data set. The increasing degree of interest in such SAR methods has arisen because of the huge numbers of compounds that may need to be synthesized and tested in the research programmes associated with the development of novel pharmaceuticals and agrochemicals. SAR methods provide a means of identifying those molecules that have a high degree of likelihood of exhibiting the desired activity, and thus permit the concentration of the synthetic programme on those areas that are most likely to bring useful results [8–12]. SAR methods may be used in two quite distinct ways. *Lead optimization* methods attempt to optimize activity within a given congeneric series by systematic modification of previously identified active compounds. *Lead generation* attempts the inherently more difficult task of identifying classes of untested compounds that might be active and that could form the basis of an optimization study.

SAR methods are based upon two fundamental assumptions. Firstly, it is assumed that there is some quantitative relationship between structural parameters and activity; a wide range of parameter types are now used including physicochemical properties, spectral characteristics, and two-dimensional or three-dimensional substructural features automatically generated from structure diagrams. The second assumption is that the relationship can be described in mathematical terms using a range of techniques including multiple regression, discriminant analysis, and supervised and unsupervised classification. This simple description of the basis of SAR methods suggests obvious practical limitations. Thus structural formulae have been developed over the years [13] primarily as a means of communicating chemical information, a role that they perform outstandingly well, rather than as a means for the comprehensive description of molecular properties. Attempts have been made to obtain such descriptions by the use of quantum mechanical methods [14, 15]. These attempts have not, however, been uniformly successful since the methods are still either inherently inaccurate or parameterized towards some specific property, and the derivation of the molecular properties from the wave equation is often difficult; moreover, the complexity of the methods is such as to severely limit their routine application on a large scale, and to encourage the use of the simpler empirical approaches discussed here. Additionally, the action of a drug is based upon a sequence of complicated physico-chemical events, each of which may depend in large part upon the nature and characteristics of all previous events. Thus while similar structures may well exhibit similar activities in some biosystem, it does not necessarily imply that structurally unrelated molecules cannot exhibit similar activities either by the same, or a different, mode of action.

At the risk of over-simplification, it is possible to divide the many SAR methods that have been described into three groups, these depending upon the processing requirements associated with an individual compound and, consequently, upon the number of molecules that may be included in an analysis. The use of quantum mechanics is restricted to very small numbers of structures owing to the complexity of the associated calculations, and similar comments apply to molecular modelling and graphical display methods. Data sets containing some tens, or a few hundreds, of structures may be processed using multivariate statistical techniques such as multiple regression or discriminant analysis; the bulk of the SAR literature refers to data sets of this size using Hansch analysis, Free–Wilson analysis or pattern recognition methods. Large-scale analysis of hundreds or thousands of compounds may be undertaken using substructural analysis.

9.1.2 Free–Wilson additivity model

This approach [16] is based on the assumption that within a series of analogues, the addition of a particular substituent group at some specified position results in a constant additive contribution to the total activity of a compound; the SAR may then be modelled by the expression

$$\text{activity} = k_1 + \sum k_{ij} X_{ij} \ .$$

In this equation, the biological activity is usually described by log $(1/C)$ where C is the concentration required to produce some standard biological response, k_1 is the overall average of activities for the whole congeneric series, k_{ij} is the contribution of the ith substituent at the jth position, and X_{ij} is an *indicator* variable that describes the presence $(X_{ij} = 1)$ or absence $(X_{ij} = 0)$ of the ith substituent at the jth position. The values of the individual contributions k_{ij} are obtained by multiple regression analysis of the observed activities of a number of analogues that have been synthesized and tested: the resulting values may then be used to predict the activity of other, untested analogues with differing combinations of substituents on the parent structure. Several modifications of the basic approach have been suggested [17, 18], while Adamson and Bawden [19] have described an extended model in which fusion points and heteroatom positions automatically derived from WLN are included in the regression analysis.

A limitation of the basic approach is that it can only be used where multiple substitution sites exist, and where there is a minimum of two occurrences for each combination of substituent and position. This implies either that a large number of analogues will need to be synthesized, or that the analysis will be restricted to a limited number of substituents or positions: these restrictions have been largely overcome in the modified model proposed by Fujita and Ban [20]. Free–Wilson analysis is frequently cited but there seem to have been relatively few applications discussed in the literature although the concept of indicator variables has now become incorporated in many applications of Hansch analysis. Perhaps the main value of the model may prove to be the concept of additive substructural contributions that underlies the class of techniques generally referred to as ***substructural analysis*** [21].

9.1.3 The Hansch multiple parameter approach

The basic assumption in this model [22], which was first introduced at about the same time as Free–Wilson analysis, is that the variations in biological activity that occur within series of compounds as a result of systematic substructural modifications may be correlated with a range of physicochemical properties: the approach might thus be more accurately described as a property–activity relationship rather than a structure–activity relationship. Hansch introduced the idea of combining several properties into a multiple parameter model in which the biological response is modelled in terms of hydrophobic, electronic, steric and dispersion components. The contribution of each of these factors is described by substituent constants that characterize the difference in some physicochemical property between the parent compound in a series, and the particular molecule that is being studied. The use of such an approach derives from the earlier work of Hammett [23] and Taft [24] on linear free energy relationships, and the electronic and steric contributions are, in fact, represented by the Hammett σ and Taft E_s values; the hydrophobic and dispersion components are generally modelled by the Hansch π and molar refractivity (MR) values. Many other parameter types have been suggested, but those listed are probably the most widely used. Of these contributions, Hansch has emphasized the importance of hydrophobic interactions in the activity of biologically active molecules, and has quantified this in terms of substituent π values. The π value of a substituent is defined as $\log(P/P_0)$, where P is the partition coefficient between octanol and water for the substituted compound, and P_0 the coefficient for the unsubstituted compound. This value is not only relatively independent of the individual series of compounds, but also may be approximated in the case of an unknown substituent by summing the Rekker hydrophobic constant values for the constituent fragments [25].

The SAR may be expressed using the formula

$$\text{activity} = k_1 + k_2\pi + k_3\sigma + k_4E_s + k_5MR$$

with the values of the coefficients k_1 to k_5 being obtained by multiple regression analysis of a data set for which both the activity and the various parameter values are known: in many cases, only one or two of the parameters may need to be included in the regression equation to obtain a good correlation. Such a linear relationship may not explain all of the variance in the data, and parabolic equations [26] or the use of indicator variables or other position-dependent parameters may need to be invoked to obtain a satisfactory correlation. However, problems of statistical validity may arise if very large numbers of variables are considered for use in the regression equation [27]. Also, reliable predictive equations are likely to be achieved only if a reasonable spread of parameter values is obtained, and several methodologies have been described to ensure that this is achieved [28–30].

Very many successful applications of Hansch analysis have now been described. It is usually, but not necessarily, limited to structural variations within a given congeneric series, where there is a high probability that the same mode of action pertains for all of the compounds: if this expectation is in large part justified, a suitable combination of parameters may usually be identified that not only gives a satisfactory correlation with the observed activities but also permits the prediction of activity in untested structures and insights to be made into the mechanism of action. Because of the widespread use of the technique, considerable effort has gone into the measurement of accurate values for the various types of substituent constant, in particular the π values, and Hansch and his coworkers have published extensive compendia of experimental data. However, there are still very many structures of interest for which the required parameters are unknown, and in such cases the values must be calculated [25, 31] or measured experimentally.

9.1.4 Pattern recognition methods

Pattern recognition (PR) methods were first used in chemistry to study the problem of assigning unknown compounds to one of a series of structural classes upon the basis of machine-readable mass spectra. This work, which has been well reviewed in the book by Jurs and Isenhour [32], led to studies using other types of spectral information and, more recently, to studies involving the prediction of biological activity in compounds [33].

There are three main stages in any pattern recognition analysis: feature extraction, training, and classification. Feature extraction involves an analysis of some set of compounds whose activities have been tested, called the *training* set, to identify a set of attributes that may be used for the characterization of the molecule; this stage may also involve the evaluation of the individual contributions of each of these attributes. Training involves the setting up of some sort of decision function that may be used to discriminate between the various activity classes present in a data set upon the basis of the attributes identified in the first stage. Finally, classification involves the assignment by the decision function of molecules whose activities are unknown to one or more of the activity classes; these untested molecules are usually referred to as the *prediction* set.

A very wide range of types of chemical descriptor has been suggested for use in PR investigations, ranging from simple atom, bond and ring counts via substructural fragments to complex three-dimensional and mass spectral descriptors [12, 34]. The main problem is that discriminating and complex substructures are generally required to give good correlations and high predictive abilities, but this necessarily implies a large number of substructural types, with two consequent problems. Firstly, the calculated activity contributions associated with a particular substructure may be based upon only a very small number of observations and, secondly, there may well be no known contributions for certain of the substructures in the molecules in the prediction set. The first of these points is of importance, owing to the fact that many statistical techniques used in PR, such as linear learning machines and discriminant analysis, require a large sample-to-feature ratio if statistical validity is to be achieved [35]. To ensure that this is so, a variety of sophisticated selection criteria have been suggested to extract some subset of the total set of attributes that retain the great bulk of the discriminatory abilities of the complete set of attributes. It is also usual to scale and normalize the observed values for the attributes if different types of attribute, such as substructures and physico-chemical properties, are to be used together in a study.

Two main training methods have been applied to chemical problems: linear learning machines and distance measures. Linear learning machines attempt to derive an $(n - 1)$-dimensional hyperplane so as to separate the n-dimensional data into groups of compounds corresponding to the observed activity classes present in the data set: thus if the biological data is simply of the active/inactive kind, one would hope to establish a hyperplane such that all of the active molecules lay to one side of the plane, and all of the inactives to the other. Classification of molecules in the prediction set may then be achieved by plotting their position in the space and determining upon which side of the hyperplane they lie. Distance measures, of which the k-nearest neighbour classification is by far the most common, are conceptually very much simpler in operation since a compound in the prediction set is classified by the majority vote of its k nearest neighbours in the training set; as a majority vote is unambiguous when the number voting is odd, k is usually set to be $1, 3, 5$ etc. The classification stage hence consists of measuring the distance between the unknown compound and all of the structures in the training set to identify the closest structures.

The classification methods described thus far are said to be *supervised* in the sense that the underlying classes present in the data are assumed to be known before classification takes place. *Unsupervised* classification, or cluster analysis, is the name given to computational methods for the identification of the groupings present in a data set. The basis for the identification of the groups, or clusters, of compounds is generally some measure of inter-molecular distance or similarity, and there has been some interest in the use of such techniques for a range of SAR applications [28, 36, 37].

Although far removed from traditional SAR methods, PR techniques have shown themselves capable of achieving impressive predictive abilities. The most extensive studies have been made by Jurs and his coworkers who have reported the development of a software package, called ADAPT, that is specifically designed for use in SAR studies involving PR methods [38]. Reports of the application of PR methods to SAR have aroused a certain degree of criticism [39–41], but experience with the methods and the availability of systems such as ADAPT should enable the implementation of the techniques on a routine basis in the future.

9.1.5 Substructural analysis methods

Substructural analysis methods are particularly well suited to large-scale screening applications in computerized chemical information systems since they are based upon the correlation of activity with the occurrence of chemical fragments, and such data may be readily generated from a machine-readable structure diagram as described in Chapter 6. The basic substructural analysis approach is described in the paper by Cramer *et al.* [21]. Given some training set for which activity data is available, an activity weight can be calculated for each of the fragment types present. A typical weight that might be calculated would be the fraction of the molecules in the training set that were active: corresponding inactivity weights can also be calculated in a similar manner. Molecules in the prediction set may then be ranked in decreasing order of expected activity by summing the weights of their constituent fragments. This simple scheme has been considerably extended by Hodes and his coworkers [42, 43] who have derived a semi-quantitative probabilistic basis for activity ranking, and have now implemented it on a routine basis as a component of the National Cancer Institute's screening programme.

Such an approach works well with nominal, that is active/inactive, property data, but is not directly suited to use with quantitative property data. This problem has been studied by Adamson and his coworkers [19, 44] who have described a modified version of Free–Wilson analysis. A regression equation is set up that relates the observed activities to individual fragment contributions; however, these contributions are assumed to be independent of position, thus removing one of the main limitations associated with traditional Free–Wilson analysis. The method has been developed specifically for use with fragment types automatically generated from either connection tables or Wiswesser Line-formula Notations, and impressive correlations have been reported for a range of types of property.

One structure-based SAR method that has aroused considerable interest in recent years is the use of topological indices – numerical indices that characterize the topology of a molecule by a consideration of the pattern of atomic interconnections. Since chemical structure diagrams may be considered as labelled graphs, many of the indices have their origins in graph theory, although information theoretic indices have also been described [45]. The most detailed studies have been made by Kier and his coworkers using molecular connectivity indices derived from a simple index introduced by Randic to provide a quantitative measure of branching in hydrocarbons [46]. This simple numerical identifier has been extensively developed with modifications being introduced to take account of heteroatoms and unsaturated bonds, and to allow for the calculation of a series of different indices corresponding to a range of substructural sizes. Although claims have been made for the non-empirical nature of molecular connectivity indices, the primary advantages are the ease with which the various types of index may be calculated, and the impressive correlations that have been reported with a wide range of types of biological activity.

9.1.6 Molecular modelling

Molecular modelling refers to the various techniques which chemists use to obtain a visual representation of chemical structures. Computer-based molecular graphics systems have found considerable application in theoretical studies of the physical and chemical properties of compounds as a function of their molecular structure [47, 48]. It is in the context of biological systems, which often involve the interaction of large and complex molecules, that the techniques are most usefully applied, and there is now considerable interest in their use for drug design [49–58].

9.1.6.1 *Mechanical and computer models*

Hand-written structure diagrams are of only limited value for representing large non-planar molecules, and provide only a poor impression of the spatial relationships between atoms. Some impression of three-dimensional perspective can be obtained in a two-dimensional drawing by use of selective shading, size gradation and 'out-of-plane' bond symbols, but perspective rapidly becomes obscured as the size and non-planarity of the molecule increases. Mechanical models provide a more tangible representation of molecular shape and volume, but these are also of limited value for large molecules, if only for practical reasons. With high-speed

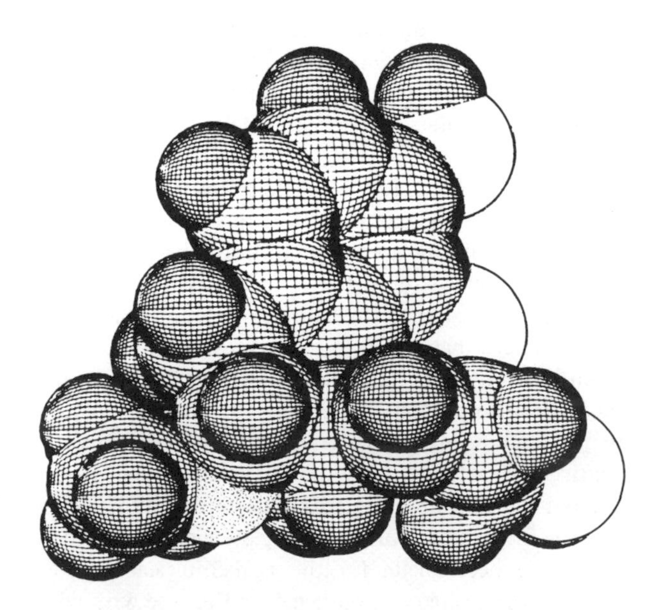

Fig. 9.1 – Space-filling projection of morphine.

computers and graphical displays, it has become possible to supplement the conventional mechanical models with computer systems which are capable of building and manipulating three-dimensional projections of molecular structures [59, 60].

Both software and hardware contribute to the overall capability of a molecular graphics system. By providing as input X-ray crystallographic data or, where these are not available, standard bond lengths, bond angles and dihedral angles, software routines can perform a variety of theoretical calculations. These calculations include free-energy minimizations and the computation of charge densities and electronic contour maps, from which accurate space-filling projections of a molecular structure can be made. Fig. 9.1 illustrates a space-filling projection of the morphine molecule, while Fig. 9.2 shows the equivalent 'ball-and-stick' model.

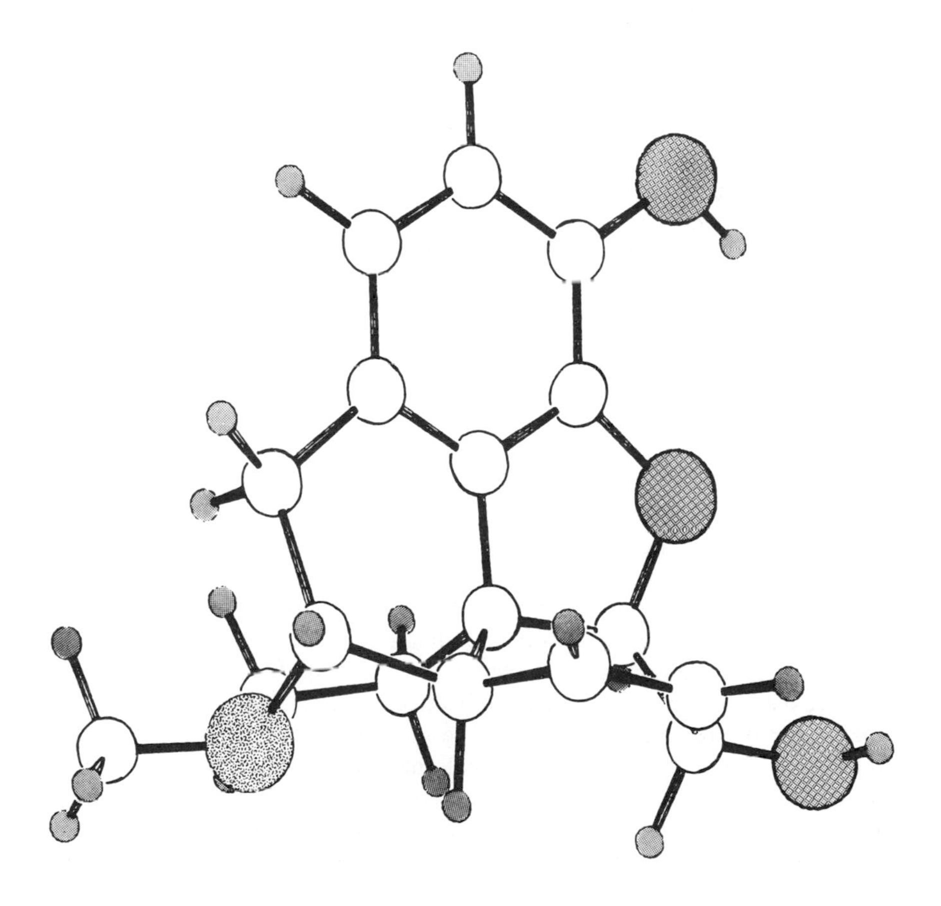

Fig. 9.2 – 'Ball-and-stick' projection of morphine.

Hardware routines, in turn, enable enhancements to the visual appearance of these projections to be made. Techniques for enhancing three-dimensional perspective include the addition of shading and colour-highlighting to atoms and bonds, the variation in the intensity of illumination across the projection to indicate depth, and the removal of 'hidden' lines. However, if the computer merely reproduces a three-dimensional projection of a chemical structure, albeit more elegantly than is possible with a mechanical model, then much of its potential power remains unused. It is the dynamic manipulation of these images that constitutes the significant advance over the use of mechanical models in providing a means of studying the interactions between molecules.

The hardware functions which enable dynamic manipulations depend upon the graphics display system used. In the more sophisticated systems, image transformations include localized rotations around selected bonds, the rotation of an entire structure projection in space, and the 'close-up' display of selected portions of a structure. In addition, several images may be displayed simultaneously and moved independently of one another on the screen, so that their relative positions and orientations are changed. Using a technique known as *plane clipping* an image can be moved in a direction perpendicular to the plane of the screen so that the molecular surfaces can be viewed from points both 'inside' and 'outside' the molecule itself. Transformations of this sort, carried out in real-time under the direction of the chemist, make it possible to visualize complex interactions of large molecules with a degree of clarity and comprehension not previously possible.

9.1.6.2 *Applications of molecular modelling to drug design*

The 'drug-receptor' theory [61] states that the biological action exerted by a drug is in part a consequence of its interaction with a receptor molecule, and that the level of biological activity is proportional to the 'goodness of fit' at the binding site between the drug and receptor molecules. Molecular modelling provides a valuable means of studying the interaction between these molecules, for example between a protein and a nucleic acid molecule. By calling software routines to adjust the shape or the orientation of a molecule, it is possible to view the effects of changing the nature of the interaction, with the result that it may be possible to propose structural modifications to an existing drug molecule in order to increase its activity or selectivity.

Since molecules interact at their surfaces, algorithms have been developed to compute and display these surfaces, and in particular that part of the molecular surface which is accessible to other molecules. Some algorithms compute a continuous surface function for a molecule from known van der Waals atomic radii, and display this in the form of several thousands of points distributed over the surface [56,60]. Each may be colour-coded and represent a single atom, and the intensity of illumination of the image may be varied in order to enhance perspective. The molecular surface of diphenylalanine computed in this way is illustrated on the cover jacket of this book. Because the entire molecular surface can be computed, interior cavities and deep clefts in the surface can be identified. Plane clipping allows the dimensions of these cavities to be revealed, and provides a remarkable means of viewing potential binding sites. These can be viewed from any chosen angle, either from positions looking 'into' the molecule, corresponding for example to the viewpoint of a drug molecule approaching its receptor site, or from positions looking 'outward' towards the approaching molecule. Simultaneous movement of the two images, guided by real-time energy minimization calculations, permits the favoured 'docking' procedures to be viewed and the mechanism of the interaction to be studied. Similarly, by calculating the steric hindrance to the approach of a molecule from specified directions and at potential reaction sites, it is possible to obtain quantitative estimates of the likelihood of certain inter-molecular reactions, and simulations of these reactions can be viewed 'as they happen' [56].

Graphics devices are described in some detail in Chapter 10, and their use in chemical structure retrieval systems is described in Chapter 7. With the continuing fall in the cost of computer memory and developments in very large scale integrated (VLSI) circuit technology (see Chapter 10) it is certain that interactive computer graphics systems, and particularly those based on raster technology, will become increasingly available for molecular modelling applications. Increases in processing speed will enable more rapid calculation of interaction energies and molecular surface projections, with the result that molecular modelling techniques will undoubtedly become extended to the investigation of molecular interactions of ever-greater complexity and interest.

9.2 COMPUTER-AIDED SYNTHESIS DESIGN

9.2.1 Introduction

The synthesis of chemical compounds forms a significant part of the cost of chemical research programmes. These costs are measured not only in terms of equipment and materials but also in terms of the time and expense devoted to the preparation and purification of large numbers of compounds, and are such that a high priority is frequently given to the elaboration of efficient, high-yield syntheses.

Two types of chemical syntheses are common. The development and testing of a drug compound will involve the synthesis of a large number of structural analogues, and these are likely to be prepared by selective transformations of a relatively small number of compounds of a similar structure to the compounds to be tested. Here the problem for the chemist is to identify for each of a series of compounds the most suitable means of effecting the necessary transformation. The second common type of chemical synthesis is the *total synthesis* of a molecule. The starting materials for a total synthesis will usually be common laboratory reagents, in which case the problem of designing the synthesis is one of identifying plausible reaction pathways which lead from small, often unspecified reagents to the synthesis of the desired *target* molecule. Here the problem is best approached from the target itself, and solutions sought from a *retro-synthetic* analysis in which reactions are applied successively in order to identify at each step a number of simpler precursor molecules. The anlaysis is considered to be complete when the precursors for the final step in the analysis correspond to available starting materials.

The elaboration of a complex synthesis can involve many hours of laboratory work. While the experienced chemist may be able to sketch the critical steps which are most likely to be involved in a successful synthesis, and to identify the types of reaction by which the necessary structural transformations may be achieved, it is seldom possible to describe fully the intermediate steps nor the order in which these must be performed. The ability of the computer to generate plausible synthetic routes is due in part to the relatively small number of known reaction types with which it can build reaction sequences, and to the selectivity with which many of these reactions proceed. The computer is able to investigate exhaustively every combination of these reactions which could lead to the synthesis of the target molecule, and can report these in an order which reflects their likelihood of success in the laboratory.

This section describes the different approaches which have been adopted in an effort to develop computer programs which assist the synthetic organic chemist in elaborating new synthetic routes [7,62–64]. The first computer-assisted synthesis design program was reported by Corey and Wipke in 1969 [65], since when details of several other computer programs have been reported. After two decades of development within academic and research institutions, some of these programs are now being used and evaluated by industry, and particularly by pharmaceutical companies in drug design programmes.

9.2.2 Approaches to automatic synthesis design

Two fundamentally different approaches underlie the existing synthesis design programs. On the one hand are those which are essentially empirical in nature. These programs make use of a library of known reactions and proceed from the target molecule in a retro-synthetic direction. Examples of these programs are LHASA (Logic and Heuristics Applied to Synthetic Analysis), SYNCHEM (Synthetic Chemistry), SECS (Simulation and Evaluation of Chemical Synthesis), and CASP (Computer-Aided Synthesis Program). LHASA [64–69] was developed at Harvard University by Corey and Wipke from an earlier prototype program OCSS (Organic Chemical Simulation of Synthesis). SYNCHEM [70–72] is a program under continuing development at the State University of New York, Stony Brook, under the direction of Gelernter. SECS [73–75] is a derivative of LHASA developed in the early 1970s by Wipke at the University of California, Santa Cruz, and conceived with an emphasis on stereochemistry which was not considered in the original LHASA program. CASP is an offshoot of SECS and is being developed in Europe by a consortium of Swiss and West German pharmaceutical companies.

The second, non-empirical approach to synthesis design is represented in programs such as EROS (Elaboration of Reactions for Organic Synthesis) [76, 77] and CAMEO [78], and in a program developed by Hendrickson at Brandeis University [79–82]. These programs do not make use of a reaction library, but instead represent structural transformations in a generalized manner by manipulating an abstract model of atoms and bonding electrons. Unlike library-based programs, which are capable of proposing reaction pathways only on the basis of known chemistry, non-empirical programs such as EROS are not constrained in this way and are able to search for synthetic routes in areas of new and unexplored chemistry.

9.2.3 Library-based synthesis design programs

When applied in the retro-synthetic direction, a structural transformation is generally referred to as a *transform*, and may represent a known mechanistic step or a single-step or multi-step reaction, as illustrated in Fig. 9.3. Programs which are based on a library of transforms proceed in the retro-synthetic direction by applying to the target molecule transforms which lead to a simplification of its structure. In subsequent steps, transforms are applied to one or more of these simplified structures in order to generate further precursor molecules. By selecting at each step only those precursors considered as suitable intermediates, the program continues to build up sequences of reactions until precursors are generated which can serve as starting materials for the proposed syntheses.

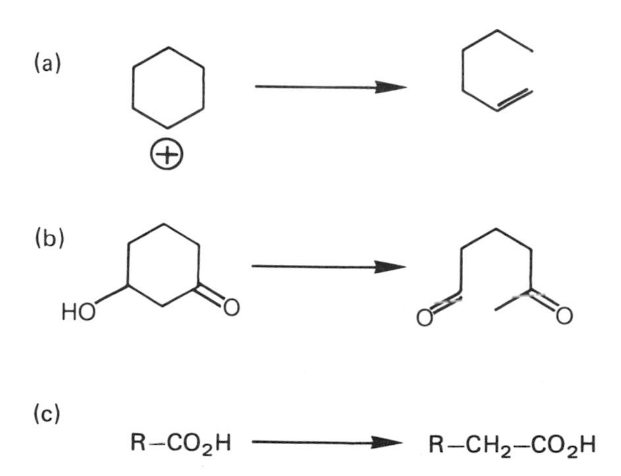

Fig. 9.3 – Structural transformations represented by (a) a mechanistic step, (b) a single-step reaction, (c) a multi-step reaction. (Reprinted with permission from *Chemical and Engineering News* May 9, 1983. Copyright 1983 American Chemical Society.)

Interactive programs such as LHASA and SECS proceed under the control of the chemist, who evaluates each of the precursors generated by the program and selects those to be processed further. SYNCHEM, on the other hand, proceeds without the intervention of the chemist; here the additional *heuristics* necessary to decide which of the precursors are suitable for further elaboration are contained within the program itself, and the output of the program consists of all of the proposed synthetic routes for the specified target molecule in the form of a *synthesis tree*, derived automatically and in its entirety.

9.2.3.1 *Transforms*

A transform is a self-contained statement of the structural changes which occur when a particular reaction is applied to a certain substructure, and of the conditions which govern the success of that reaction. A transform library may consist of some hundreds or thousands of transforms, and forms a part of the *knowledge-base* associated with these programs. A design principle which is characteristic of all rule-based expert systems and to which all synthesis design programs based on a transform library adhere is that modifications or additions to the knowledge-base should not entail any changes to be made to the program itself. Similarly, a transform is written in such a way that it does not reference any other transform, with the result that transforms can be added to or deleted from the knowledge-base without altering existing transforms. Special languages have been devised for encoding transforms: CHMTRN (Chemistry Translator) is a language used with LHASA, while ALCHEM (Associative Language for Chemistry) has been devised for use with SECS.

Transforms used by LHASA and by SECS are similar in nature, and contain the following information:

(1) Reference details, including the transform number and name, and literature citations to publications in which details of the reaction can be found; a minimal description of the structural changes which result from the reaction; and an initial rating value which quantifies the synthetic usefulness of that transform.

(2) Conditional expressions, or *qualifiers*, which, when evaluated in the context of a target structure, enable the program to decide on the suitability of applying the transform. Evaluating the qualifiers may cause the initial rating value to be reduced, or may eliminate the transform outright.

(3) Reaction conditions which are necessary for the reaction to take place, including temperature, solvents, catalysts and co-reagents.

(4) A generalized description of the mechanism by which the structural transformation takes place, and which the program uses to generate the precursor.

Fig. 9.4 illustrates the LHASA transform for the alkylation of an acid halide. The qualifiers, rating modifications, experimental conditions and reaction mechanism are readily understood from the CHMTRN description.

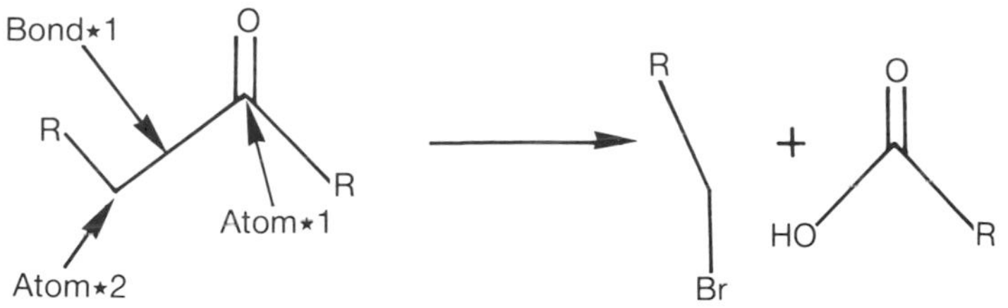

```
TRANSFORM 269
Name Alkylation of Acid Halide
.. C–C=O ⟶ CX + HOOC–
.. March 362; B+P 716; TET LETT 4647 (1970), 2113 (1971)
.. Path Change 0
Rating 30
Subject Group is Ketone
Object Group is Acid
Disconnective

Kill if there is a Leaving Group on Atom*2
Kill if Bond*1 is in a Ring ... Not Intramolecular
Subtract 10 if there is no Hydrogen on Atom*2 ... Harder RXN
Subtract 30 if Atom*2 is a Bridgehead ... Harder RXN
Subtract 50 if Alpha to Atom*2 is a Leaving Group
      ... Grignard Eliminates
Subtract 30 if there is a Functional Group & Anywhere on Path
      ... Interfering Group
Subtract 30 if Atom*2 is a Stereocenter

Conditions in Fragment*2 RMGX and in & Fragment*1 SOC12

Attach an Alcohol to Atom*1
Attach a Bromide to Atom*2
Break Bond*1
```

Fig. 9.4 – LHASA transform for alkylation of acid halides. (Reprinted with permission from *Chemical and Engineering News* May 9, 1983. Copyright 1983 American Chemical Society.)

The qualifiers embody the heuristics which enable these programs to direct the search to the most plausible syntheses. The heuristics express in a concise and formalized way the existing chemical knowledge, and each is expressed in terms which reflect its generality of application. By evaluating the heuristics, the program decides which actions should be taken in certain circumstances and which should be avoided. In the case of interactive programs, such as LHASA and SECS, the most powerful heuristics, which are often those least susceptible to formalization, are supplied by the chemist by interaction with the program. In this way, the chemist evaluates the proposed reaction sequences at each stage of the analysis, and directs the operation of the program by selecting those sequences to be explored further. Interactive design programs therefore benefit from the ability of the user to evaluate, and if necessary to eliminate, reaction sequences at the earliest opportunity, thereby constraining the familiar 'combinatorial explosion' which is characteristic of problems of this nature.

In contrast to interactive programs, SYNCHEM attempts to discover synthetic routes without online guidance, and its success in achieving this has established the feasibility of non-interactive synthesis discovery by computer. For this purpose, SYNCHEM uses additional heuristics to those embodied in its reaction library, and these enable the program to select those of its proposed reaction pathways which are suitable for further exploration. For example, one such heuristic estimates the 'cost' of reaching the target molecule from the current intermediate, measured in terms of the cumulated yield and the ratings of each of the participating reactions. A further heuristic estimates the difficulty of synthesizing that intermediate from available starting materials, and predicts the probable yield of the intermediate along the best path from the starting materials. This is possible since SYNCHEM is capable of operating in the *synthetic* as well as in the retro-synthetic direction, and is able to generate the products of a given set of reactants for a specified sequence of reactions. The starting materials recognized by SYNCHEM presently comprise a subset of the Aldrich catalogue of commercially available organic compounds.

9.2.3.2 *Perception and strategy control*

The generation of plausible syntheses is a complex problem, and can only be solved efficiently by establishing and then following a carefully conceived strategy. The chemist identifies suitable approaches to the synthesis of a compound intuitively, drawing on his experience and knowledge of synthetic principles. Synthesis design programs are capable of formulating strategies automatically [73]. Here, an initial strategy is formulated by perceiving in the target molecule the structural features which are known to be of synthetic importance, for example isolated, bridged and fused rings, functional groups, strategic bonds and stereo-centres. From these the program is able to decide upon the most likely approaches to the synthesis of the target molecule.

In addition to automatic strategy formulation, LHASA and SECS offer a number of 'standard' short-range and long-range strategies for common structural transformations, and the chemist is able to choose from among these in order to direct the program to develop a particular synthetic approach. The strategies and options available in LHASA are shown in Table 9.1.

Table 9.1 – LHASA retro-synthesis strategies and options.

Group-oriented strategy	Bond-oriented strategy
Opportunistic	Bridged strategic
Disconnective	Fused strategic
Reconnective	Appendages – Ring appendage only
Unmasking	– Branch appendage only
	Manual designation

Long-range strategy	Options
Stereospecific C=C	One-group only
Diels–Alder	Two-group only
Robinson annulation	Pattern only
Small rings	Preserved bonds
Halolactonization	Internal protection
Quinone Diels–Alder	Stereosimplifying
Birch reduction	

There are three major strategies for retro-synthetic analysis. These are sub structure or *group-oriented* strategies, *bond-oriented* strategies, and *long-range* strategies based on multi-step reaction pathways. Each major strategy comprises a number of sub-strategies, or *tactics*, and the chemist may invoke these in order to direct the way in which the synthesis is developed. Group-oriented strategies include *disconnective*, *reconnective* and *unmasking* tactics. Disconnective tactics permit only the use of transforms which break a carbon–carbon bond, while reconnective tactics use transforms which cause an increase in cyclic connectivity in the synthetic direction. By invoking the unmasking tactic, the program uses only those transforms which unmask functionality, for example by removal of protecting groups.

When a bond-oriented strategy is selected, only transforms which disconnect the specified *strategic* bond are attempted. Bond-oriented tactics include disconnections of bonds between cyclic and acyclic components of a molecule, and the breaking of bonds in bridged- and fused-ring systems in order to simplify the cyclic network of the molecule [83]. In all of these cases, strategic bonds are identified automatically. The tactic of *manual designation* enables the chemist himself to specify a set of strategic bonds, in which case only transforms which disconnect one or more of these are attempted.

Applying a strategy to a molecule generates a list of *goals*, often expressed in terms of the structural changes which must be carried out in order to realize that strategy. Transforms are selected within a given strategy in an attempt to achieve these goals, and in the case where no single transform can achieve a particular goal, that goal is automatically resolved into a number of *subgoals*. Transforms are then selected in order to achieve each subgoal. The selection of transforms for a particular goal or subgoal is made by matching the substructures or strategic bonds which are required by each transform with the significant features perceived in the target molecule. The transforms are then applied, and a new set of precursor molecules is generated from which new goal lists are constructed.

The long-range strategy in LHASA offers the chemist a choice of a number of general reaction types which are of established synthetic importance, for example annulation, oxidation and reduction reactions. Here the chemist chooses the type of strategy which he considers to be the most appropriate method of synthesizing the desired target molecule, and the program automatically generates the goals and subgoals which must be achieved before that strategy can be applied. Unlike the group-oriented and bond-oriented strategies, LHASA processes long-range strategies with a minimum of interaction with the chemist. Fig. 9.5 illustrates a simple long-range strategy based on a Diels–Alder route to the synthesis of 4-methylhydroxy methylcyclohexane (I), and involves only two subgoal transformations [64]. It is not possible to apply the Diels–Alder transform directly to the target molecule because the corresponding synthetic reaction is known always to result in an unsaturated cycle by virtue of the diene reactant. Furthermore, the Diels–Alder cyclization reaction proceeds more readily if there is an electron-withdrawing group on the dienophilic reactant. LHASA recognizes that a hydroxy-substituted dienophile, suggested by the target structure, is not suitable for this type of reaction, and establishes a subgoal intermediate (III) in which a methyl ester group replaces the primary alcohol functionality. The program automatically generates the subgoal intermediate (III) by applying two transforms, the first of which introduces the double bond into the cyclohexane ring (II), and the second of which replaces the alcohol functionality by the methyl ester group. Both of these steps correspond to reductions in the forward, synthetic direction. Having attained the subgoal intermediate (III), the Diels–Alder transform is applied and leads to the two simple molecules methylacrylate (IV) and 2-methyl-1,3-butadiene (V). The chemist recognizes each of these as available in the laboratory and terminates the program at this point.

9.2.3.3 *Display of synthetic routes*

One option for the display of results from synthesis design programs is in the form of a *synthesis tree*, in which each node represents a molecule generated by the program and in which nodes are linked together into the proposed synthetic pathways. In the case of SYNCHEM, the synthesis tree can be displayed and individual synthetic intermediates and reaction pathways examined only at the

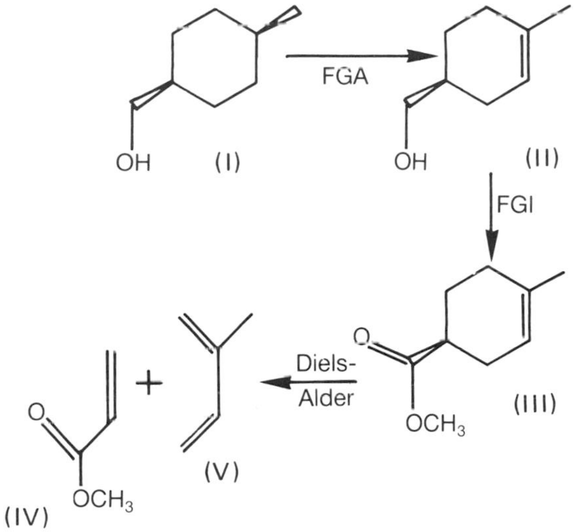

FGA Functional Group Addition
FGI Functional Group Interchange

Fig. 9.5 – Diels–Alder route to the synthesis of 4-methylhydroxy methylcyclo hexane. (Reprinted with permission from *Chemical and Engineering News* May 9, 1983. Copyright 1983 American Chemical Society.)

end of the processing. With LHASA and SECS, precursors can be inspected as they are generated by the program, and the current synthesis tree can be displayed at any stage during the processing. LHASA displays a synthesis tree in the form illustrated in Fig. 9.6. This synthesis tree has been generated by applying the Robinson annulation strategy to the target molecule shown [84]. The synthesis tree displays the target molecule and each of its synthetic precursors, and identifies the suggested starting materials at the end of each branch. Each molecule in the tree is numbered uniquely: a number at the head of a pair of dashed lines identifies a precursor which consists of more than one molecular fragment, and the dashed lines lead to the individual molecules involved.

A variety of display options permits the chemist to examine a particular molecule and each of its precursors, and to inspect complete routes from the target molecule to starting materials. For example, by selecting the 'Get-Family' option and pointing by appropriate means to structure *17* (Fig. 9.6), the precursors *18*, *19* and *20* are displayed. Choosing the 'Get-Lineage' option and pointing to structure *18* displays the entire synthetic route from starting material *18* through precursors *17*, *16*, *15* and *14* to the target molecule *2*.

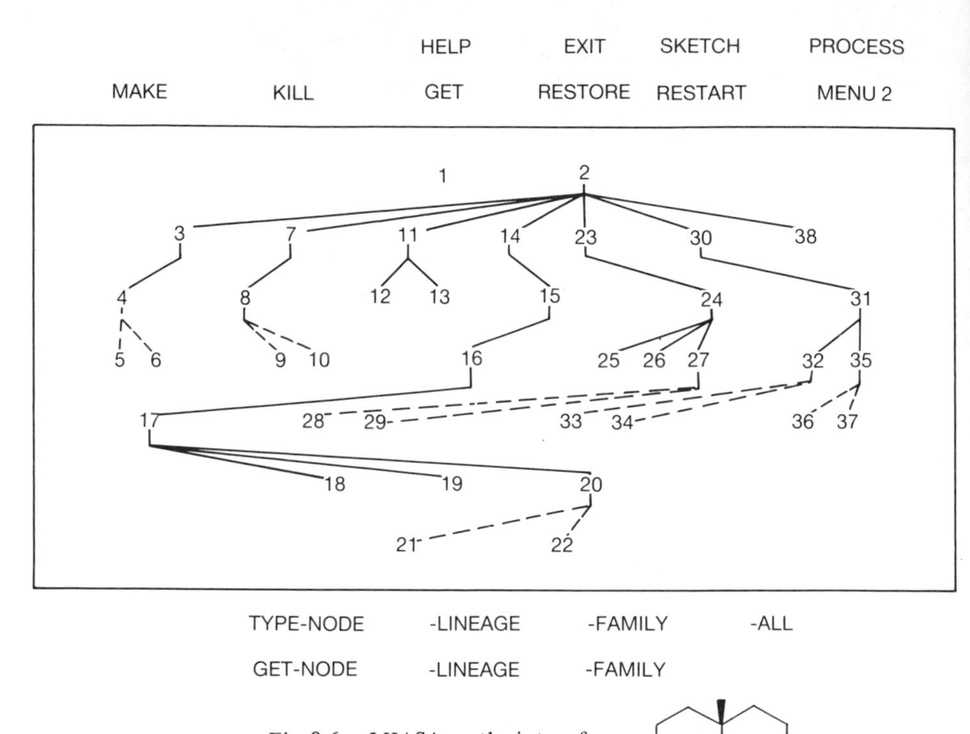

Fig. 9.6 – LHASA synthesis tree for

9.2.4 Non-empirical approaches to synthesis design

Non-empirical approaches to chemical synthesis design tend at present to be of greater academic than practical interest, and have not been evaluated as thoroughly as programs such as LHASA, SECS and SYNCHEM. Two approaches are described here in outline only – that embodied in the EROS program, and that due to Hendrickson.

The EROS program does not use a reaction library to elaborate chemical synthesis, and has the capability of proceeding in either the synthetic or the retro-synthetic direction, depending on the nature of the synthesis problem. In the scheme a chemical reaction is considered as the conversion of one *molecular ensemble* (*EM*), that is the set of reactant molecules, into an isomeric *EM* which represents the products of the reaction. Each of the reactant and product molecules is represented in the form of a 'bond-electron' *BE*-matrix, in which the rows and columns correspond to individual atoms of the molecule. The reaction itself is represented by a further matrix, the *R*-matrix, which describes the redistribution of valence electrons which occurs during the reaction. A more complete description of these representations is given in Chapter 8.

By identifying reaction sites in the target molecule, in terms of the bonds which are susceptible to being broken, and then mapping the *R*-matrices onto the *BE*-matrix of the target, a number of new *BE*-matrices is produced each of which represents a possible precursor molecule. The matrix operations which redistribute the electrons in order to produce each new *BE*-matrix are perfomed without a knowledge of any particular chemical reaction, with the result that the structural transformations may correspond to known chemistry or may be completely novel. The *BE*-matrix of each precursor molecule may then, in turn, be operated upon by the *R*-matrices.

For each precursor, EROS calculates the approximate reaction energy which is needed for that particular transformation in the synthetic direction, and the precursors are ranked on this basis. Output from EROS is similar to that of SECS and, in addition to the display of two-dimensional structure diagrams, includes the facility for computing and displaying three-dimensional structure projections.

The emphasis of Hendrickson's approach to chemical synthesis design is to seek efficient synthetic routes by converging as rapidly as possible on available starting materials. The basic strategy is to reduce to a minimum the number of intermediates which must be generated to complete a synthesis. This is achieved by seeking whole routes between starting materials and the target rather than by generating and evaluating intermediates one step at a time. This philosophy differs from that of programs such as LHASA, SECS and SYNCHEM, in which synthetic routes are developed by selecting precursors at each level of the synthesis tree. Analysis of the problem in Hendrickson's program is based on the rationale that the synthesis of a chemical molecule consists primarily of linking together smaller molecules, with the emphasis on construction reactions as the means of achieving this. Optimization of a synthetic route depends on minimizing the number of steps involved and not on factors such as yield predictions and energy considerations. The shortest, and in this context the optimal, path will not include steps in which changes in functionality are required. Accordingly, the functionality resulting from any single construction step is required to be exactly that which is necessary for the subsequent step.

The synthesis problem is subdivided by identifying from a list of available starting materials all of the possible combinations of reactants from which the specified target molecule may be constructed. Each set of reactants will give rise to one possible synthesis of the target molecule. Each complete synthesis is
represented in the form of a bondset, which describes
represented in the form of a *bondset*, which describes the structure of the reactant and product molecules and identifies the atoms in the reactant molecules at which the necessary construction reactions must occur in order to produce the target. The construction reactions are stored in a reaction library. Unlike transforms, each reaction is represented as two independent *half-reactions*, where each half-reaction defines for one of the reactant molecules the changes in the

functionality from reactant to product. A *polarity* is associated with each half-reaction, and is based on whether the entities are nucleophilic or electrophilic; an intermolecular construction reaction must combine two half-reactions of opposite polarity. For each bondset, one or more sequences of half-reactions is defined which satisfy the initial and final functionalities required by the bondset. The shortest sequence of half-reactions is identified for each bondset, and these are displayed as the group of optimal synthetic routes.

As in the case of EROS, the proposed syntheses are not constrained by known chemistry. This is a consequence of the construction reactions being described as independent half-reactions. Despite this, Hendrickson's protocol is currently somewhat limited in that it is restricted to construction reactions, and is not capable of developing synthetic routes which include cleavage and rearrangement reactions.

9.2.5 Future developments in chemical synthesis design

In the case of each of the empirical programs LHASA, SECS and SYNCHEM, there is a need for more comprehensive reaction libraries which cover a wider field of chemistry. For example, LHASA is at present strong in aliphatic chemistry, and in particular the total synthesis of natural products, but is rather poor in the areas of heterocyclic and aromatic chemistry. Industrial users of LHASA in the UK and the USA are contributing substantially to the expansion of the transform database by addition of new transforms in these areas. The CASP program now operates with a database of some 4000 transforms.

In the case of LHASA, expansion of its knowledge-base to include heterocyclic and aromatic chemistry has resulted in the need for more elaborate means of matching the substructural features of the target molecule with those of each transform. LHASA classifies and uses transforms on the basis of functional groups, which makes it difficult to utilize, for example, reactions which create heterocyclic ring systems. A research group at the University of Leeds is at present extending the *pattern-matching* algorithms of LHASA so that transforms may be 'keyed' on arbitrary substructures rather than simply on arrangements of functional groups.

Other important extensions to LHASA include the recognition of molecular symmetry during the perception and strategy planning stages, and the greater utilization of stereochemistry in exploring synthetic pathways. Since its inception, SECS has been particularly strong in its utilization of stereochemical information, using the calculation of spatial relationships and steric effects in the evaluation of transforms. Further developments of both LHASA and SECS involve implementation of a bidirectional approach to the generation of reaction sequences, as embodied in SYNCHEM. The ability to search in the synthetic direction as well as in the retro-synthetic direction can be of considerable advantage where a specific starting material or class of structural goals can be identified.

Even when such developments are incorporated, problems may still remain. Firstly, there is generally little provision made for the routine updating of the transform library or for the modification of the transform ratings in the light of operational experience. Secondly, the libraries tend to contain only reactions of proven and general synthetic value, which are expressed in terms which maximize their applicability over as wide a range of structural types as possible. Many reactions that have been reported in the literature will not be included in the transform library because of their lack of generality of application, even though they may well proceed in excellent yield under certain circumstances.

The scope and power of synthesis programs could be dramatically increased if access to such reactions was possible, and a means by which this might be achieved has been suggested [85, 86]. The approach makes use of the fact that reaction indexing and synthesis design programs are complementary in nature, with reaction indexing programs being used for the efficient production of relatively simple characterizations of very large numbers of reactions (see Chapter 8), while synthesis design programs, conversely, carry out highly sophisticated manipulations on the much smaller number of reactions in the transform library.

A target molecule is input to the synthesis design program and potential syntheses are generated using a transform library in the manner described above. Each synthesis proposed by the program will contain one or more substructural transformations, and each transformation may be used as a query to search a reaction index containing as many reactions as possible. Since a given transformation may usually be brought about in several ways, it is clear that very many more possible synthesis routes will become available, with each being based upon the same set of substructural changes. Such facilities have been built into the REACCS reaction indexing system described in Chapter 8 to allow an interface with synthesis design progams.

Such an approach is particularly useful in the case of programs like EROS which are not based on a transform library and which operate by making and breaking bonds in the target molecule. The transformations resulting from these changes may, or may not, correspond to known chemical reactions: in the latter case, of course, the proposed synthetic route may not be chemically feasible. An external reaction index would be essential in evaluating the merit of the large number of routes output by such a program since the chemist would need only consider those reaction pathways that did, in fact, correspond to known chemical behaviour. In addition, a reaction index might also focus attention on those steps in an otherwise useful synthetic sequence for which new chemistry must be explored.

REFERENCES

[1] Haggin, J., Computers Shift Chemistry to More Mathematical Basis, *Chem. Eng. News*, 7–20, May 9 (1983).

[2] Nilsson, N. J., *Principles of Artificial Intelligence*, Tioga, Palo Alto, Calif. (1980).
[3] Nau, D. S., Expert Computer Systems, *Computer*, 63–85, Feb. (1983).
[4] Davis, R., Buchanan, B. G., and Shortliffe, E. H., Production Rules as a Representation for a Knowledge-Based Consultation Program, *Artificial Intelligence*, **8**, 15–45 (1977).
[5] Lindsay, R. K., Buchanan, B. G., Feigenbaum, E. A., and Lederberg, J., *Applications of Artificial Intelligence for Organic Chemistry*, McGraw-Hill, New York (1980).
[6] Smith, D. H. (ed.), *Computer-Assisted Structural Elucidation*, ACS Symposium Series Vol. 54, American Chemical Society, Washington D.C. (1977).
[7] Wipke, W. T., and Howe, W. J., (eds.), *Computer-Assisted Organic Synthesis*, ACS Symposium Series Vol. 61, American Chemical Society, Washington D.C. (1977).
[8] Topliss, J. G. (ed.), *Quantitative Structure–Activity Relationships of Drugs*, Academic Press, New York (1983).
[9] Redl, G., Cramer, R. D., and Berkoff, C. E., Quantitative Drug Design, *Chem. Soc. Rev.*, **3**, 273–292 (1974).
[10] Cramer, R. D., Quantitative Drug Design, *Ann. Rep. Med. Chem.*, **11**, 301–310 (1976).
[11] Martin, Y. C., A Practitioner's Perspective of the Role of Quantitative Structure–Activity Analysis in Medicinal Chemistry, *J. Med. Chem.*, **24**, 229–237 (1981).
[12] Bawden, D., Computerized Chemical Structure-handling Techniques in Structure–Activity Studies and Molecular Property Prediction, *J. Chem. Inf. Comput. Sci.*, **23**, 14–22 (1983).
[13] Rouvray, D. H., The Changing Role of the Symbol in the Evolution of Chemical Notation, *Endeavour*, **1**, 23–31 (1977).
[14] Kier, L. B., *Molecular Orbital Theory in Drug Research*, Academic Press, New York (1971).
[15] Christoffersen, R. E., Quantum Pharmacology: Recent Progress and Current Status, *ACS Symposium Series*, **112**, 3–19 (1979).
[16] Free, S. M., and Wilson, J. W., A Mathematical Contribution to Structure–Activity Studies, *J. Med. Chem.*, 7, 395–399 (1964).
[17] Cammarata, A., Interrelationship of the Regression Models used for Structure–Activity Analysis, *J. Med. Chem.*, **15**, 573–577 (1972).
[18] Kubinyi, H., and Kehrhahm, O. H., Quantitative Structure–Activity Relationships. I. The Modified Free–Wilson Approach, *J. Med. Chem.*, **19**, 578–586 (1976).
[19] Adamson, G. W., and Bawden, D., A Substructural Analysis Method for Structure–Activity Correlation of Heterocyclic Compounds using Wiswesser Line Notation, *J. Chem. Inf. Comput. Sci.*, **17**, 164–171 (1977).

[20] Fujita, T., and Ban, T., Structure–Activity Study of Phenethylamines as Substrates of Biosynthetic Enzymes of Sympathetic Transmitters, *J. Med. Chem.*, **14**, 148–152 (1971).

[21] Cramer, R. D., Redl, G., and Berkoff, C. E., Substructural Analysis. A Novel Approach to the Problem of Drug Design, *J. Med. Chem.*, **17**, 533–535 (1974).

[22] Hansch, C., A Quantitative Approach to Biological Structure–Activity Relationships, *Acc. Chem. Res.*, **2**, 232–239 (1969).

[23] Hammett, L. P., The Effect of Structure upon the Reactions of Organic Compounds. Benzene Derivatives, *J. Amer. Chem. Soc.*, **59**, 96–103 (1937).

[24] Taft, R. W., Separation of Polar, Steric and Resonance Effects in Reactivity, in M. S. Newman (ed.), *Steric Effects in Organic Chemistry*, Wiley, New York (1956), pp. 556–675.

[25] Rekker, R. F., *The Hydrophobic Fragment Constant*, Elsevier, Amsterdam (1977).

[26] Hansch, C., and Clayton, J. M., Lipophilic Character and Biological Activity of Drugs. II. The Parabolic Case, *J. Pharm. Sci.*, **62**, 1–21 (1973).

[27] Topliss, J. G., and Edwards, R. P., Chance Factors in QSAR Studies, *ACS Symposium Series*, **112**, 131–145 (1979).

[28] Hansch, C., Unger, S. H., and Forsythe, A. B., Strategy in Drug Design. Cluster Analysis as an Aid in the Selection of Substituents, *J. Med. Chem.*, **16**, 1217–1222 (1973).

[29] Wootton, R., Cranfield, R., Sheppey, G. C., and Goodford, P. J., Physicochemical–Activity Relationships in Practice. 2. Rational Selection of Benzenoid Substituents, *J. Med. Chem.*, **18**, 607–613 (1975).

[30] Streich, W. J., Dove, S., and Franke, R., On the Rational Selection of Test Series. I. Principal Component Method Combined with Multidimensional Mapping, *J. Med. Chem.*, **23**, 1452–1456 (1980).

[31] Chou, J. T., and Jurs, P. C., Computer-assisted Computation of Partition Coefficients for Molecular Structures using Fragment Constants, *J. Chem. Inf. Comput. Sci.*, **19**, 172–178 (1979).

[32] Jurs, P. C., and Isenhour, T. L., *Chemical Applications of Pattern Recognition*, Wiley, New York (1975).

[33] Stuper, A. J., Brugger, W. E., and Jurs, P. C., *Computer-Assisted Studies of Chemical Structure and Biological Function*, Wiley, New York (1979).

[34] Brugger, W. E., Stuper, A. J., and Jurs, P. C., Generation of Descriptors from Molecular Structure, *J. Chem. Inf. Comput. Sci.*, **16**, 105–110 (1976).

[35] Stuper, A. J., and Jurs, P. C., Reliability of Non-parametric Linear Classifiers, *J. Chem. Inf. Comput. Sci.*, **16**, 238–241 (1976).

[36] Adamson, G. W., and Bush, J. A., A Method for the Automatic Classification of Chemical Structures, *Inform. Stor. Retr.*, **9**, 561–568 (1973).

[37] Rubin, V., and Willett, P., A Comparison of some Hierarchal Monothetic Divisive Clustering Algorithms for Structure Property Correlation, *Anal. Chim. Acta*, **151**, 161–166 (1983).

[38] Stuper, A. J., and Jurs, P. C., ADAPT: A Computer System for Automated Data Analysis using Pattern Recognition Techniques, *J. Chem. Inf. Comput. Sci.*, **16**, 99–105 (1976).

[39] Perrin, C. L., Testing of Computer-assisted Methods for Classification of Pharmacological Activity, *Science*, **183**, 551–552 (1974).

[40] Matthews, R. J., A Comment on Structure–Activity Correlations Obtained using Pattern Recognition Methods, *J. Amer. Chem. Soc.*, **97**, 935–936 (1975).

[41] Wold, S., and Dunn, W. J., Multivariate Quantitative Structure–Activity Relationships (QSAR): Conditions for their Applicability, *J. Chem. Inf. Comput. Sci.*, **23**, 6–13 (1983).

[42] Hodes, L., Hazard, G. F., Geran, R. I., and Richman, S., A Statistical Heuristic Method for Automated Selection of Drugs for Screening, *J. Med. Chem.*, **20**, 469–475 (1977).

[43] Hodes, L., Computer-aided Selection of Compounds for Antitumour Screening: Validation of a Statistical–Heuristic Method, *J. Chem. Inf. Comput. Sci.*, **21**, 128–132 (1981).

[44] Adamson, G. W., and Bush, J. A., Method for Relating the Structure and Properties of Chemical Compounds, *Nature*, **248**, 406–407 (1974).

[45] Kier, L. B., and Hall, L. H., *Molecular Connectivity in Chemistry and Drug Research*, Academic Press, New York (1976).

[46] Randic, M., On Characterization of Molecular Branching, *J. Amer. Chem. Soc.*, **97**, 6609–6615 (1975).

[47] Marshall, G. R., Bosshard, H. E., and Ellis, R. A., Computer Modeling of Chemical Structures: Applications in Crystallography, Conformational Analysis, and Drug Design, in Wipke, W. T., Heller, S. R., Feldmann, R. J., and Hyde, E. (eds.), *Computer Representation and Manipulation of Chemical Information*, Wiley, New York (1974).

[48] Feldmann, R. J., The Design of Computing Systems for Molecular Modeling, *Ann. Rev. Biophys. Bioeng.*, **5**, 477–510 (1976).

[49] Levinthal, C., Molecular Model-building by Computer, *Sci. Am.*, **214**, 42–52 (1966).

[50] Katz, L., and Levinthal, C., Interactive Computer Graphics and Representation of Complex Biological Structures, *Ann. Rev. Biophys. Bioeng.*, **1**, 465–504 (1972).

[51] Feldmann, R. J., Bacon, C. R. T., and Cohen, J. S., Versatile Interactive Graphics Display System for Molecular Modelling by Computer, *Nature*, **244**, 113–115 (1973).

[52] Gund, P., Three-dimensional Pharmacophoric Pattern Searching, *Progress in Molecular and Subcellular Biology*, **5**, 117–143 (1977).

[53] Feldmann, R. J., Bing, D. H., Furie, B. C., and Furie, B., Interactive Computer Surface Graphics Approach to Study of the Active Site of Bovine Trypsin, *Proc. Natl. Acad. Sci.*, **75**, 5409–5412 (1978).

[54] Gund, P., Andose, J. D., Rhodes, J. B., and Smith, G. M., Three-dimensional Molecular Modeling and Drug Design, *Science*, **208**, 1425–1431 (1980).

[55] Fox, J. L., Computer Graphics Aid Study of Molecules, *Chem. Eng. News*, 27–29, July 21 (1980).

[56] Langridge, R., Ferrin, T. E., Kuntz, I. D., and Connolly, M. L., Real-time Color Graphics in Studies of Molecular Interactions, *Science*, **211**, 661–666 (1981).

[57] Humblet, C., and Marshall, G. R., Three-Dimensional Computer Modeling as an Aid to Drug Design, *Drug Development Research*, **1**, 409–434 (1981).

[58] Morffew, A. J., Bibliography for Molecular Graphics, *J. Molecular Graphics*, **1**, 17–23 (1983).

[59] Smith, G. M., and Gund, P., Computer-generated Space-filling Molecular Models, *J. Chem. Inf. Comput. Sci.*, **18**, 207–210 (1978).

[60] Max, N. L., Computer Representation of Molecular Surfaces, *IEEE Trans. Comput. Graph. Applic.*, **3**, 21–29 (1983).

[61] Danielli, J. F., Moran, J. F., and Triggle, D. J., (eds.), *Fundamental Concepts in Drug–Receptor Interactions*, Academic Press, New York (1970).

[62] Bersohn, M., and Esack, A., Computers and Organic Syntheses, *Chem. Rev.*, **76**, 269–282 (1976).

[63] Gasteiger, J., Computer-assisted Synthesis Design. Present State and Future Perspectives, *Chim. Ind. (Milan)*, **64**, 714–721 (1982).

[64] Long, A. K., Rubenstein, S. D., and Joncas, L. J., A Computer Program for Organic Synthesis, *Chem. Eng. News*, 22–30, May 9 (1983).

[65] Corey, E. J., and Wipke, W. T., Computer-assisted Design of Complex Organic Syntheses, *Science*, **166**, 178–192 (1969).

[66] Corey, E. J., Wipke, W. T., Cramer, R. D., and Howe, W. J., Computer-assisted Synthetic Analysis. Facile Man–Machine Communication of Chemical Structure by Interactive Computer Graphics, *J. Amer. Chem. Soc.*, **94**, 421–430 (1972).

[67] Corey, E. J., Wipke, W. T., Cramer, R. D., and Howe, W. J., Techniques for Perception by a Computer of Synthetically-significant Structural Features in Complex Molecules, *J. Amer. Chem. Soc.*, **94**, 431–439 (1972).

[68] Corey, E. J., Cramer, R. D., and Howe, W. J., Computer-assisted Synthetic Analysis for Complex Molecules. Methods and Procedures for Machine Generation of Synthetic Intermediates, *J. Amer. Chem. Soc.*, **94**, 440–459 (1972).

[69] Pensak, D. A., and Corey, E. J., LHASA – Logic and Heuristics Applied to Synthetic Analysis, *ACS Symposium Series*, **61**, 1–32 (1977).

[70] Gelernter, H. L., Sridharan, N. S., Hart, A. J., Yen, S. C., Fowler, F. W., and

Shue, H., The Discovery of Organic Synthesis Routes by Computer, *Top. Curr. Chem.*, **41**, 113–150 (1973).

[71] Gelernter, H. L., Sanders, A. F., Larsen, D. L., Agarwal, K. K., Bovie, R. H., Spritzer, G. A., and Searleman, J. E., Empirical Explorations of SYNCHEM, *Science*, **197**, 1041–1049 (1977).

[72] Agarwal, K. K., Larsen, D. L., and Gelernter, H. L., Applications of Chemical Transformations in SYNCHEM2, a Computer Program for Organic Synthesis Route Discovery, *Computers and Chemistry*, **2**, 75–84 (1978).

[73] Wipke, W. T., Braun, H., Smith, G., Choplin, F., and Sieber, W., SECS – Simulation and Evaluation of Chemical Synthesis: Strategy and Planning, in ref. [7] pp. 97–127.

[74] Wipke, W. T., Ouchi, G. I., and Krishnan, S., Simulation and Evaluation of Chemical Synthesis – SECS, *Artificial Intelligence*, **11**, 173–193 (1978).

[75] Wipke, W. T., and Dyott, T. M., Simulation and Evaluation of Chemical Synthesis. Computer Representation of Stereochemistry, *J. Amer. Chem. Soc.*, **96**, 4825–4834 (1974).

[76] Gasteiger, J., and Jochum, C., EROS, a Computer Program for Generating Sequences of Reactions, *Top. Curr. Chem.*, **74**, 93–126 (1978).

[77] Gasteiger, J., Jochum, C., Marsili, M., and Thoma, J., The Synthesis Planning Program EROS, *MATCH*, **6**, 177–199 (1979).

[78] Salatin, T. D., and Jorgensen, W. L., Computer-assisted Mechanistic Evaluation of Organic Reactions: Overview, *J. Org. Chem.*, **45**, 2043–2051 (1980).

[79] Hendrickson, J. B., Systematic Synthesis Design. The Scope of the Problem, *J. Amer. Chem. Soc.*, **97**, 5765–5784 (1975).

[80] Hendrickson, J. B., A Systematic Organization of Synthetic Reactions, *J. Chem. Inf. Comput. Sci.*, **19**, 129–136 (1979).

[81] Hendrickson, J. B., and Braun-Keller, E., Systematic Synthesis Design. Generation of Reaction Sequences, *J. Comp. Chem.*, **1**, 323–333 (1980).

[82] Hendrickson, J. B., Braun-Keller, E., and Toczko, G. A., A Logic for Synthesis Design, *Tetrahedron, Suppl.* **9**, 359–370 (1981).

[83] Corey, E. J., and Jorgensen, W. L., Computer-assisted Synthetic Analysis. Synthetic Strategies Based on Appendages and the Use of Reconnective Transforms, *J. Amer. Chem. Soc.*, **98**, 189–203 (1976).

[84] Corey, E. J., Johnson, A. P., and Long, A. K., Computer-assisted Synthetic Analysis. Techniques for Efficient Long-range Retrosynthetic Searches Applied to the Robinson Annulation Process, *J. Org. Chem.*, **45**, 2051–2057 (1980).

[85] Gund, P., Andose, J. D., and Rhodes, J. B., Computer-assisted Synthetic Analysis in Drug Research, in ref. [8] pp. 179–187.

[86] Willett, P., The Evaluation of an Automatically Indexed, Machine-readable Chemical Reactions File, *J. Chem. Inf. Comput. Sci.*, **20**, 93–96 (1980).

10

Developments in computing

10.1 INTRODUCTION

The rapid development of computer techniques over the last few years has had a profound influence on the design and use of information systems in general, and of systems for the provision of chemical information in particular. Indeed, the move from manual to computerized retrieval systems is evidenced by the change in name of the main research journal in the field from *Journal of Chemical Documentation* to *Journal of Chemical Information and Computer Sciences.*

Computer-based chemical information systems are probably better developed, and offer more sophisticated means of retrieval, than systems in almost any other area of research. This is due to the availability not only of bibliographic and numeric data but also of the chemical structure diagram as a simple and flexible machine-readable representation for structural information, and previous chapters have discussed some of the many ways in which such information may be processed, stored and retrieved. This chapter seeks to provide a background by describing some of the technologies that underlie these applications and systems, and to give an overview of some of the current developments in computing and telecommunications: it should be emphasized that the choice of topics is by no means exhaustive, but has been made to highlight those developments that are most likely to affect the designers and users of chemical information systems.

The chapter commences with a description of computer hardware, this being the component of computing systems that is most apparent to users. After considering the characteristics of processors and digital storage devices, attention is focused on a range of input and output devices, in particular those which seek to provide a user-friendly interface so that systems may be operated by end-users, rather than by intermediaries such as computer or information specialists. Particular attention is paid to speech handling and to chemical graphics, which increasingly form the basis for the structure handling systems described in previous chapters. Computer software for some application may be obtained either by writing the required routines in some programming language or by the purchase of an available package, and these two alternatives are discussed in the second section of the chapter. The final section deals with telecommunication systems, concentrating upon new telecommunications media and the local area networks that will form the basis of much distributed computing in the future.

10.2 DEVELOPMENTS IN HARDWARE TECHNOLOGY

10.2.1 Processor characteristics

The traditional classification of computers as mainframe, mini or micro is becoming increasingly difficult with the introduction of 'super' versions of each, and with the constantly increasing levels of performance made possible by developments in silicon technology [1].

Supercomputers have generally been developed in response to the demands of extremely large-scale computation that are characteristic of fields such as meteorology, reactor design and image processing. Supercomputers are characterized [2,3] by large main stores, high bandwidth channels to backing storage devices, and facilities for extremely fast scalar and vector processing: typical of such computers is the Cray-1 which has a semiconductor main store of one million 64-bit words with single-bit error correction and double-bit error detection and a cycle time of only 12.5 nanoseconds [4]. Rather than attempting to further optimize the speed and capacity of a single processor, a considerable amount of interest is now being expressed in the use of array processors. These consist of a number of processors operating in parallel under the control of an instruction unit, and such machines are thus typical of the class of computers known as Single Instruction stream, Multiple Data stream (SIMD). Nearly all current computers, conversely, are based on Single Instruction stream, Single Data stream (SISD) architectures in which instructions are processed in sequence upon individual data elements from the main store. A typical SIMD computer is the International Computers Limited (ICL) Distributed Array Processor (DAP) [5] which is based upon a 64 × 64 (4096) matrix of processing elements, each of which has an associated 16k bit store. The DAP is currently implemented as part of the main store of a large ICL 2900 host mainframe and a DAP installation is thus, like all supercomputers, extremely expensive; however, ICL have recently announced that a 32 × 32 DAP will shortly be available attached to a dedicated workstation, the total cost being comparable with that of a small minicomputer system.

Although offering lower levels of performance than supercomputers, and based upon SISD architectures, modern-day mainframes are still capable of extremely high processing rates for a wide range of computational loads. Thus machines such as the International Business Machines (IBM) 308X and Amdahl 580 series involve the extensive use of large-scale integration (LSI) and very large-scale integration (VLSI) components and large cache memories that permit processing rates in the range 10–20 millions of instructions per second. There is a noticeable trend to use tightly coupled multiprocessor systems in which several computers are linked together to give an improved performance over that obtainable from a single machine. However, it has been found that a law of diminishing returns applies, with each additional processor adding less and less to the overall performance of the system; in addition, extremely complex systems software is required for the control of such multiprocessor systems.

Early minicomputers, such as the Digital Equipment Corporation (DEC) PDP-11 series, were characterized by limited 16-bit memories, but these are now being replaced in superminis, such as the VAX-11/780 or Gould SEL 32/87, by 32-bit wordlength main stores and virtual storage operating systems that are capable of supporting the large numbers of interactive terminals characteristic of academic and research environments. Conventional minicomputers are now increasingly being challenged by the new generation of 16-bit microcomputers which offer, for example, 512 kbyte main stores, 20 Mbyte Winchester disc units and sophisticated graphics capability, and currently retail in the UK at a few thousand pounds. These machines, such as the Sirius or the IBM Personal Computer, are often basically 8-bit processors that have been configured to permit 16-bit operation, but more powerful microcomputers, which feature multi-user operating systems and true 16-bit operation, are beginning to appear in large numbers and it is likely that such machines will replace many current 16-bit minicomputers over the next few years. Currently, the great bulk of micros are based upon 8-bit wordlengths, and developments at this level are increasingly aimed at reducing costs, as is perhaps most strikingly exemplified by the Sinclair ZX-80 and ZX-81, and by improving the range and quality of the software available. The microcomputer market is currently in a state of flux with further improvements in the sophistication and performance of silicon technology being accompanied by the arrival in the market place of machines from traditional computer manufacturers such as IBM, ICL and DEC: it seems possible that many of the small microcomputer manufacturers will merge, or disappear, within the next few years.

The capabilities of computers depend in large part upon the performance characteristics of the components that are used for the construction of the central processing unit; at present, these are based upon silicon products, but there is now a considerable amount of interest in alternative technologies that could be used to form the basis for future processor types. Superconducting computers [6], based upon the principle of the Josephson junction effect, have been extensively studied and offer the prospect of sub-nanosecond switching times. However, superconductivity manifests itself only at very low temperatures, and thus any practical computer based upon this technology would need to operate at temperatures close to absolute zero: indeed, several designs assume the embedding of the processor in liquid helium to reduce temperatures sufficiently. An alternative approach is based upon the use of gallium arsenide, and similar Group III/Group V substances, which appear to offer comparable switching speeds to superconducting devices but without the requirement for very low operating temperatures. More futuristic, and still faster, designs include optical computers and biochips which are based upon laser and molecular switching respectively. However, computers based upon such approaches are likely to remain at the research stage for several years yet.

10.2.2 Backing storage devices

When we come to consider backing storage devices, rather than central processors, magnetic tapes still have a role to play in computing systems as an archival storage medium, but the continuing demand for interactive computing systems means that discs, of one sort or another, now form the most important type of backing storage device. Whether considering floppies for microcomputers or thin film fixed discs for mainframes, the trend is for constantly increasing storage densities, although the dramatic increases in capacity that have been achieved over the last few years have not in general, been accompanied by a comparable decrease in access times, and thus solid state backing stores may well have an increasingly important role in the future.

Perhaps the most important development in this area is that of optical disc technology [7,8] which has derived in large part from the video discs used in the home entertainments industry. Perhaps the most widely publicized example is the Philips system in which data is digitally encoded as a series of microscopic pits that are burnt into the surface of a tellurium-coated plastic disc by a laser. The data may then be read by accurately focusing a laser onto the surface: if no pit has been burnt, the beam will be reflected and may be detected using a mirror system. Data is recorded on 40,000 tracks per side giving a total capacity of about 2,000 Mbyte per disc at a cost which is already less than modern magnetic disc units such as the comparably sized IBM 3380; by building up disc packs, it is expected that this capacity will be increased by one or two orders of magnitude within the very near future. There are still some problems associated with the provision of facilities for the high-speed positioning of the laser beam to permit random access to the data, but it should surely prove possible to switch a light beam considerably faster than it is possible to move the read and write heads in current magnetic disc stores.

A limitation is that, at present, a disc cannot be rewritten; however, the potential cheapness of the technology means that this is not likely to be a problem for those application areas that do not require frequent updating of the data. One such area is that of online bibliographical retrieval systems where the search files are updated only at fairly regular, well-spaced intervals and where there is now a move away from short, keyword-based document representatives to the storage of the full texts of documents. This facility has been available in certain specialized areas, such as computerized legal retrieval [9], for some years now; but the exceedingly low costs associated with optical discs mean that this type of retrieval may soon be extended to the major scientific and technological databases. Apart from full text, the optical disc also allows the provision of local graphical information to supplement textual data retrieved from an external source: this approach forms the basis for the Video PATSEARCH system [10] of Pergamon InfoLine in which patents retrieved from an existing online text file may be used to access a local disc that contains the associated diagrams and drawings. Such mixed media systems are likely to become increasingly popular in the near future, at least until digital encoding or facsimile technology [11] develop sufficiently to permit the economic transmission of large amounts of graphical information over telecommunications channels.

10.2.3 Input and output technologies

As far as the user of computing systems is concerned, some of the most striking developments are taking place in the means by which data is entered into and extracted from computers. Punched cards and paper tape have given way as input devices to electronic keyboards, touch-sensitive screens, light pens and digitizing tablets, while simple lineprinter output is being supplemented by high resolution colour graphics and the use of high quality printing devices. In the longer term, speech looks set to play an increasingly important role for both input and output.

Optical character recognition (OCR) machines have been available for many years now, but it is only recently that they have become sufficiently reliable in operation to provide relatively error-free means of generating machine-readable data from printed media. Perhaps the most innovative example of such a device is the Kurzweil Data Entry Machine. This was originally designed as a reading machine for the blind with a sensitive OCR unit being coupled to a speech synthesizer to provide a remarkably effective, albeit expensive, means of providing access to printed media for the blind without the need to use Braille. The recognition programs have now been considerably extended and the Kurzweil machine is being widely used as a general-purpose data input medium [12].

However, the most common input medium is via a keyboard of some sort, and the spread of word processing systems has focused attention on possible improvements in a device that has remained almost unchanged for many decades. The size of keyboards has increased with the proliferation of special function keys, while the actual shape and design are being increasingly affected by ergonomic considerations of various kinds. Perhaps the greatest potential change is in the physical layout of the keys since alterations to the familiar QWERTY arrangement may permit considerable increases in keyboarding speeds; however, while this would be of use to those who make extensive use of word processors, the great majority of keyboard users might well object to having to learn such a new arrangement.

For many purposes, the combination of menu-based software and touch-sensitive visual display unit (VDU) screens provides a fast means of data entry, and one that is particularly attractive to those unfamiliar with keyboards. Touch-sensitive screens in combination with light pens also form an increasingly important component of interactive computer graphics systems for applications such as chemical substructure searching and computer-aided design. An alternative technology involves the use of a digitizing tablet in which a human-controlled cursor or joystick is used to address locations on a VDU screen: typical of these systems is the so-called mouse which is used to control access to the Rank Xerox 8010 Star and ICL PERQ workstations and to the Apple LISA executive micro-computer.

VDU screens and paper form the most important output devices and this seems likely to be true for several years yet since the proliferation of VDUs associated with the 'office of the future' has done little or nothing to hasten the advent of the 'paperless society'. Indeed, new and improved types of printer are becoming available to meet a range of user needs. Thus the resolution of dot matrix printers has increased sufficiently to make them competitive in quality with many daisy wheel printers, while demands for large volumes of high quality output have led to the introduction of laser printers, which can produce an entire page of output at a time [13]. There has also been a revival of interest in the use of ink jet printers in which an electromagnetic field is used to focus a jet of ink into the desired shape [14]; such printers can now produce letter quality text as well as permitting the printing of graphics and chemical structures and formulae. Visual display units currently use cathode ray tubes (CRTs) but these are very bulky if portability or large displays are required. These problems may be overcome with the development of flat screen CRTs and liquid crystal devices.

10.2.3.1 *Computerized handling of speech*

Perhaps the most fascinating input–output technology is that of speech. Speech synthesis techniques have been around for quite some time, but early synthesizers essentially digitized a speech signal for subsequent reproduction, rather than actually generating the signal as required. As well as being more flexible in operation, true synthesis methods are less demanding of computer storage and are not restricted to the Dalek-like tones of many early machines. The most obvious example of modern synthesis technology is the Speak and Spell toy introduced by Texas Instruments in 1978, in which children are asked in quite natural tones to key in the spelling for some word and the resulting input is then compared with an internal dictionary of correct spellings. This seemingly trivial, but highly profitable, application has opened the way to the development of sophisticated speech synthesis chips for a wide range of tasks, such as toys, car and aircraft alarm systems, automatic telephone answering and the Kurzweil machine mentioned above. While the actual quality of the speech produced may bear some improvement, speech as an output medium would seem to be well established.

A much more challenging problem is the use of speech as an input medium, and work in this area is progressing much more slowly [15]. Individual word recognition requires the spoken input to be analysed and then compared to word patterns known to the machine until the best matching pattern is identified. There are many complications involved in this seemingly simple task, such as the minimal acoustic differences between many words or the problems resulting from background noises, variant pronounciations and room acoustics. However, recognition accuracies in excess of 95 per cent may be routinely obtained, even in quite noisy environments, as long as the size of the vocabulary is not too large. A vocabulary of one or two hundred words might appear to be unduly limiting but is, in fact, quite sufficient for most immediate application areas, such as bibliographic retrieval [16], password or telephone number identification, air traffic control, or voice activation of domestic appliances or robots. More difficult is the problem of connected speech recognition where some means must be found of breaking the continuous input signal down into a set of basic units, such as phonemes, and then building these back into complete words and hence into sentences; this must be achieved, moreover, even when many of the phoneme identifications are incorrect. Intense developmental work over some ten years has, to date, resulted in systems that offer success rates of about 70 per cent for the recognition of sentences based on vocabularies of up to 100 words: however, it has not proved generally possible to achieve such high rates with a range of speakers unless excessive training is undertaken. An example of the current state of the art is given by Erman *et al.* [17] in their description of the HEARSAY-II system developed at Carnegie Mellon University. Generalized speech understanding systems that could handle relatively unrestricted dialogue from a large population of speakers in unconstrained environments would appear to lie still many years in the future; however, as noted above, this need not delay the wide-scale introduction of speech recognition in more limited circumstances.

10.2.3.2 *Computer graphics systems*

The most important means of communication for the chemist is, of course, via the chemical structure diagram and much attention has accordingly been paid to the application of computer graphics packages [18] to the input and output of structural information at a computer terminal [19–22].

Until the end of the 1960s most commercially available graphics hardware was limited to the production of output only, using devices such as pen plotters or CRTs. Moreover, these were often expensive and required complex hardware and software interfaces to their host computer, which was usually some sort of mainframe system. A major breakthrough in the availability and ease of use of computer graphics came with the introduction of the Tektronix 401X series of storage displays, a series that is still in widespread use today and that is catered for in most of the commonly used chemical graphics software packages

[19]. Such devices are relatively inexpensive, and can be used remotely over serial telecommunication links. They introduced the important concept of hardware independence in which, in theory at least, a terminal can be connected to any type of host computer so long as the manufacturers provide an appropriate graphics library or driver.

Modern graphics terminals may be based upon either calligraphic (vector drawing) or raster (scanning) technologies. Calligraphic displays can draw a line from any point on the screen to any other, and since they are drawn by analogue circuitry the resolution of a calligraphic display is in principle infinite; however, in practice, the resolution depends on how closely two vectors can be drawn without one obscuring the other. A typical modern display for use in chemical applications has a resolution of 1024*1024, allowing 1 million distinct vector endpoints to be drawn, although very much greater resolutions may be required in other application areas such as that of chip design. A problem associated with the provision of high resolution vector displays is the need to provide extremely high data transfer rates between the host processor used for the generation of the display and the graphics device which is to draw it. This problem becomes particularly apparent in applications where the information on the screen needs to be changed in a continuous manner, as for example during the rotation of a molecule in real time as described in Chapter 9. In order to produce a display that is free of flicker, and to give an illusion of continuous movement, a screen refresh rate of approximately 1/30th second is desirable. Where dynamic display involves the redrawing of several hundreds of vectors on each update, data transfer rates of the order of 1 million bits/sec are necessary. For these reasons, there is an increasing tendency to use minicomputers as dedicated graphics processors; however, even with the increasing power of such systems, high performance calligraphic displays are still expensive, especially when full colour is required, and are likely to remain so within the near future.

The dramatic fall in the costs of computer memory and processing has recently made raster technology attractive for many applications. Raster displays are not based on vector graphics, but on a display comprising a large number of individually addressable picture units, or pixels. The information displayed on the screen is stored in a refresh memory, which is necessarily large for modern displays which support colour, variable intensity, and picture planes. Picture planes correspond to partitions of the refresh memory, each capable of storing a single screen of information. Dynamic transformations, such as three-dimensional rotation, are made possible on raster devices by alternate display of picture planes, some being refilled with updated information while others are transferred from the refresh memory to the display unit. However, dynamic processing is somewhat slower than with calligraphic devices, and at present no commercially available raster system matches the high performance calligraphic processors in this respect. Nevertheless, raster systems are somewhat less expensive than calligraphic displays since the actual display device is much simpler, and it is likely that as VLSI circuit technology is developed and hardware costs are further reduced, interactive raster graphics systems of comparable resolution and speed will become available for applications such as the work on molecular modelling that is described in Chapter 9.

Both raster and calligraphic devices can be used in animation mode where pictures are computed offline at less than the refresh rate, and then loaded into different picture planes as required; an illusion of movement may be obtained by rapidly flicking between these planes.

Rather than merely providing display facilities, there is an increasing tendency to the use of interactive graphics stations which additionally provide graphical input, i.e. the ability to control and modify the display interactively, and screen hardcopy, so that a user can obtain a permanent record of a session at the terminal both cheaply and quickly. Graphics input can be controlled by a number of devices such as thumbwheels or potentiometer knobs, which are particularly useful for entering continuously varying quantities such as the orientation of a molecule or the value of a parameter like a chemical shift, or by means of joysticks which offer either two or three degrees of freedom, and light pens or styluses and tablets, these being especially suited to the input of structural information since a rough, free-hand drawing may be converted by software to a high quality display. In many systems, only the endpoints of bonds need to be drawn, and atomic symbols, individual rings and common ring systems can be selected from a menu and placed in the appropriate position on the screen as illustrated in Chapter 6 in the context of chemical structure retrieval systems. The provision of hardcopy output is an important feature of a graphics facility. Two main approaches have been suggested for providing a record of what has taken place at a terminal: these are to duplicate all calls which plot to the screen by forming a separate plot file and then plotting it offline, or by dumping all the information on the screen. The latter screendump approach has several advantages and is becoming quite widely available, especially with raster devices.

Extended descriptions of computer graphics systems, although not in the chemical context, are given in a recent text [18].

10.3 DEVELOPMENTS IN SOFTWARE TECHNOLOGY

10.3.1 Introduction

Developments in software technology have been much less marked than in the case of hardware, and while hardware costs continue to fall rapidly, software costs have risen dramatically over the last few years and now often constitute the largest item of expenditure in the budget of a computer installation. Research into natural language processing, logical calculus, program verification and structured programming methodologies *inter alia* has had some influence on the design and implementation of software, but much is as it was ten years ago. In particular, COBOL and FORTRAN (in one form or another) maintain their traditional pre-eminence for data processing and scientific computation respectively, while many mainframe computers still operate under the control of operating systems that had their genesis in the 1960s. Changes are, however, beginning to take place.

10.3.2 Programming languages and operating systems

New, or improved, programming languages [23] are constantly being introduced for use in particular application areas, and some of these languages may become of considerable importance in the future, such as ADA, which is discussed below, the logic programming language, PROLOG, which is being considered as one of the main implementation vehicles for the software components of fifth-generation computer systems [24], and LOGO, which has aroused considerable interest as a means of introducing computers to young children. At the same time, traditional languages are enjoying a new lease of life as they are implemented on microcomputer systems: thus the MicroFocus Cis-COBOL compiler has received widespread recognition, while interpreted or compiled BASIC is the standard language for nearly all current micros. Educationalists have strongly criticized the widespread use of BASIC owing to its non-modular and unstructured character. In response to this, many versions of the language have been introduced that include DO . . WHILE and similar constructions, and it seems likely that BASIC's grip on the lowest levels of the computer market will remain for several years at least. Similar developments have been made in FORTRAN, so that FORTRAN 77 is quite modular in design.

A language that has grown dramatically in importance over the last few years is PASCAL [25]. This was originally designed for teaching the principles of structured programming but the efficiency, power and flexibility of the language, coupled with its ease of implementation, have led to it becoming a standard language in academic computing in general, as well as a major rival to BASIC as a microcomputer language. In the latter respect, it has been considerably aided by the widespread use of the p-System developed at the University of California, San Diego, to facilitate PASCAL programming on a wide range of microcomputer types. The p-System, which is itself written in PASCAL, incorporates both the language and an operating system, and produces object code, called p-code, for a hypothetical processor called a p-machine. A real microprocessor may then execute the p-code by writing a machine language emulator so that the microprocessor appears to be a p-machine; this technique means that the system can be easily implemented on a very wide range of computer types, and it represents one of the most successful attempts to date to achieve true software portability.

Perhaps the most important development in the spread of programming standards has come with the introduction of the language ADA [26]. This was developed at the behest of the United States Department of Defense as a means of halting the spread of incompatible languages, of reducing the life cycle maintenance costs of software, and of increasing the reliability of the embedded real-time computing systems that are characteristic of missile and avionics systems. The language definition arose as the result of a detailed study of the characteristics of many current high-level languages, and is, at first appearance, most obviously derived from PASCAL. However, ADA is not just a programming language since the specification also details the software tools and operating environment in which ADA programs are to be implemented; the full definition is wide ranging and this has occasioned several criticisms of the language on the grounds of ambiguity and complexity [27]. However, the widespread enthusiasm that has greeted the language suggests that it will play an increasingly important role in large-scale computing systems over the next twenty years.

The desire for software standards has also resulted in the development of simple and flexible operating systems that can be easily implemented on a wide range of computer types. This is particularly true of microcomputer operating systems where CP/M has become a *de facto* standard, and a similar process would now seem to be in progress for the UNIX operating system as a basis for supermicro and minicomputer systems, and also for academic computing [28]. It is interesting to note that UNIX was originally developed by just two research workers at Bell Laboratories in 1969–70, whereas the operating system written by IBM for its System/360 series of computers, OS/360, required more than 5,000 man-years of development effort [29].

10.3.3 Software packages

While these developments will undoubtedly influence software systems in the future, current users must make the best of what is available and there are two alternative means of acquiring applications programs: the design and implementation of software in-house, or its purchase from an external software vendor. In the latter case, a further two possibilities exist, with the software either being built to fit the user's specifications, or being purchased as a program package. Both of these approaches have certain strengths and weaknesses.

Purpose-built software has the advantage that it should, in theory, give an efficient and effective solution to the problem in hand since it has been designed for this specific purpose. Against this must be considered the design, development, testing and debugging costs, and the need to maintain the programs so as to encompass changes in user needs and in the hardware upon which they are implemented.

With the increasing costs of software development, the use of program packages becomes more and more attractive, and the number of packages available on the market grows steadily, even for fairly recondite applications. Moreover the costs of packages for mainframe and minicomputers have remained relatively stable while those for micros, which are already low, are becoming steadily cheaper as the size of the market grows by leaps and bounds; in addition, packages for micros generally have significantly lower capabilities than the corresponding mini and mainframe software, and this factor again helps to reduce costs. Apart from the matter of cost, the use of a package should permit the immediate implementation of a working, error-free system, and the availability of support if problems arise or user requirements change. It would thus seem that packages can provide the answer to most problems but in practice there are several drawbacks which, perhaps, do not become evident without a considered evaluation of what is available, and just what tasks are to be performed. The first such drawback is that good software, i.e. software that is reliable, flexible and efficient, is extremely expensive and time-consuming to develop, requiring man-years of effort for all but the smallest packages; accordingly, the purchaser of software should be aware that the quality of the product is closely related to the price that is asked. Moreover, the general lack of staff with software expertise means that much of the software that is available is of poor quality, and many package vendors are incapable of providing adequate customer support. Secondly, the fact that a package is offered as a general-purpose solution implies that it may not be ideal for a specific application, a fact that may not become fully apparent until it has been installed: this is particularly true if modifications are needed to allow for interactions with the host operating system and other applications programs. Thirdly, the development of software is dominated and restricted by a lack of suitable standards, and even where such standards exist they are not fully adopted. Finally, and perhaps most importantly, true machine independence remains a dream with all but the most limited and inefficient packages being linked to a particular machine, or requiring extensive modifications if new hardware is introduced. The firm link between hardware, the language in which an application is developed, and its subsequent performance is likely to remain for many years yet. In spite of all of these problem areas, it may nevertheless still be attractive to consider purchasing a package, as this is often the only way an objective can be met at a reasonable cost, and within a reasonable timescale.

Chemical information services are likely to acquire several types of software package to support the research and development activities of their parent organizations. Apart from packages for handling structural data, which are discussed elsewhere in this volume, the most important are likely to be database management systems (DBMS) [30,31] and free text retrieval systems (FTX) [32] for the storage and retrieval of internally generated data. Although having many features in common [33–35], there are noticeable differences between these two types of package since DBMSs are used for manipulation of formatted numerical data while FTXs operate upon the unstructured text of documents.

DBMS technology is well established in commercial and business data processing environments but their generality of approach means that problems may be encountered when they are applied to the huge amounts of heterogeneous data accumulated in many chemical projects. Thus both McTaggart and Radcliffe [36] and Ravenscroft and Smith [37] found it necessary to develop sophisticated systems for the storage of biological and clinical test data owing to the lack of suitable packages on the market. FTX packages are used for the storage and retrieval of items such as journal articles, technical reports, memoranda etc.; systems offer facilitates for database creation, update and search, and may also make provision for SDI and for the production of printed indexes. Program packages such as ASSASSIN, CAIRS and STATUS are well-established and their prevalence is likely to increase as more internally generated data is captured in machine readable form with the adoption of word processing on a large scale.

10.4 DEVELOPMENTS IN DATA COMMUNICATIONS TECHNOLOGY

10.4.1 Introduction

The national and international public-switched telephone networks (PSTN) were developed to support analogue voice signals and it is thus hardly surprising that many problems became evident when they first began to be used for data communication purposes. Most of the technical problems have now been overcome, and data transmission occupies a constantly increasing fraction of channel usage. It is not intended to discuss here the various types of information service that may be expected to result from these developments; instead, the reader is referred to the excellent review by Raitt [38]. In addition, the extent to which such new services become available may be dictated as much by political and regulatory considerations [39] as by technological innovations.

Two factors in particular have hastened the growth of data communication networks. The first is the widespread emergence of digital switching and telecommunications systems which means that there is no longer a need to convert between analogue and digital signal forms when transmitting information over a distance, and voice, text, data and graphical information may all be represented in a single form for transmission purposes. Secondly, the conventional PSTN circuit switching technology has been complemented by packet switching techniques which can deal much more effectively and economically with the discontinuous nature of much data traffic by sharing line utilization, as well as allowing significantly lower error rates and the implementation of a very wide range of interfaces to different types of computing equipment [40].

10.4.2 Transmission media

Telecommunication transmission media fall easily and naturally into two groups: connected line media such as coaxial cable or glass fibres, and free space media through which radio waves may be transmitted or broadcast. The most common type of line channel is the simple twisted wire pair but these offer too small a bandwidth and are too error-prone to form the basis of any sort of reliable data transmission system. Higher data transmission rates can be accommodated by coaxial cable which forms the basic transmission medium for long-haul national telephone trunk routes: thus a single cable can support over 10,000 voice channels and the bandwidth is sufficient to accommodate not only voice and data but also textual and graphic information, albeit at a rather low rate in some cases. It is broadband coaxial cable that is likely to form the basis for the extensive recabling that will be needed for the provision of national cable TV and videotex services. However, an increasing body of opinion suggests that optical fibre technology, in which information is transmitted in digital form by pulses from a laser or light-emitting diode, should be the preferred medium for the recabling owing to the rapid increase in the cost of copper which means that cable is now little cheaper than glass fibre while offering lower levels of performance in many areas. The advantages of glass fibre for data transmission systems are the massive potential bandwidth, which is already of the order of several hundreds of Mbaud, coupled with extremely low attenuation losses that reduce the need for repeaters, and a freedom from electrical interference that results in very low error rates. While the technology is by no means fully developed yet, it seems likely that fibre systems will form the basis for many new data transmission systems over the next decade. In the UK, a consortium of Barclays Bank, Cable and Wireless, and British Petroleum has set up the Mercury network, which is the first network licensed to compete with British Telecom (BT) following legislation to remove the monopoly of the latter. Mercury will involve the linking of most of the major cities by fibre optic cables laid alongside the rail network so as to reduce line installation costs; the services to be offered include not only conventional data processing but also document delivery and video conferencing.

The potential of communications satellites was first perceived by Arthur C. Clarke in 1945 [41] and the satellites launched for the International Telecommunications Satellite Organization (INTELSAT) [42] now provide the great bulk of international telecommunication links. These links were originally designed primarily for voice purposes, but the huge bandwidths that are now possible means that they are also eminently suitable for video links, such as trans-Atlantic teleconferencing, and data processing applications, though considerations of data protection may well do much to regulate the amount of trans-national data flow that does take place. The current generation of INTELSAT satellites are capable of supporting 13,400 telephone circuits and two television channels, and order of magnitude increases in capacity may be expected in the near future. As well as the decrease in costs arising from increased channel capacity, further economies may be expected owing to changes in the relative costs of ground stations and satellite launches on the one hand, and installation and maintenance costs associated with ground-based cables on the other. These economies have been recognized by the introduction of satellite-based data transmission facilities for private businesses, and the many recent announcements of direct satellite broadcasting systems.

10.4.3 Telecommunication networks

The most obvious manifestation of telecommunications technology on the information community is ready access to national and international online bibliographical retrieval systems via commercial networks such as Tymnet or Telenet, or the inter-governmental Euronet [43]. These networks are now well established, and developments in this area are primarily concerned with enhancements to the services available, rather than dramatic changes in the basic technology.

The main current area of interest is local area networks (LAN). These are networks that connect computing hardware within a limited geographical area, as with a set of word processors, Winchester discs and shared printers within a building, or the many VDUs etc. connected to a central minicomputer on a research site [44]. The reason for this interest is the fact that about 60 per cent of all business equipment communication takes place within one building or a small complex of buildings, 32 per cent over distances of up to 500 miles, and a mere 8 per cent over greater distances. The primary requirements of a LAN are that it should permit high data transmission rates at a low cost, while still permitting a high degree of system reliability. Probably the most successful example is the Ethernet system that was originally developed to support the research activities of the Xerox Palo Alto laboratories, but which is now forming the basis for a plethora of office automation product announcements following the publication of a joint standard by Xerox, DEC and Intel in November 1980 [45]. However, this specification has not yet been accepted as an industry standard, and many organizations within the UK and Europe have adopted ring networks in which each of the components in the network is connected into a closed loop around which messages circulate. This design reflects the extensive work carried out at the University of Cambridge where numerous peripherals, microcomputers and mainframes have been interconnected by a simple high speed interface [46]. Ethernet, conversely, is based upon a bus configuration in which components are connected to a single length of cable, this approach having the advantage that devices can be attached to, or removed from, the network without affecting any of the other components that are attached. IBM, meanwhile, has announced an extensive development program to design VLSI chips for a local network based on yet another standard, and it will obviously be some time before the goal of open systems interconnection becomes a reality.

REFERENCES

[1] Noyce, R. N., Microelectronics, *Sci. Am.*, **237**, 63–69 (1977).
[2] Kozdrowicki, E. W., and Theis, D. J., Second Generation Vector Supercomputers, *Computer*, **13**, 71–83 (1980).
[3] Levine, R. D., Supercomputers, *Sci. Am.*, **246**, 112–125 (1982).
[4] Russell, R. M., The CRAY-1 Computer System, *Comm. ACM*, **21**, 63–72 (1978).

[5] Gostick, R. W., Software and Hardware Technology for the ICL Distributed Array Processor, *Aust. Comp. J.*, **13**, 1–6 (1981).

[6] Matisoo, J., The Superconducting Computer, *Sci. Am.*, **242**, 38–53 (1980).

[7] Sigel, E., Schubin, M., and Merrill, P., *Videodiscs: the Technology, the Application, and the Future*, Knowledge Industry, White Plains (1981).

[8] Goldstein, C. M., Optical Disc Technology and Information, *Science*, **215**, 862–868 (1982).

[9] Bull, G., A Brief Survey of Developments in Computerised Legal Information Retrieval, *Program*, **15**, 109–119 (1981).

[10] Schulmann, J. L., Video PATSEARCH: a Mixed-Media System, *Inform. Technol. Lib.*, **1**, 150–156 (1982).

[11] Cawkell, A. E., *An Investigation of Commercially Available Facsimile Systems*, British Library Research and Development Department report no. 5719 (1982).

[12] Jennings, P. G., Newman, L. E., and Wilkinson, W. B., Data Capture by Optical Scanning of Published Material for Database Enhancement, *Program*, **16**, 17–26 (1982).

[13] Keen, A. J., Advanced Technology in Printing: the Laser Printer, *ICL Tech. J.*, **1**, 172–179 (1979).

[14] Kuhn, L., and Myers, R. A., Ink-jet Printing, *Sci. Am.*, **240**, 120–132 (1979).

[15] Reddy, D. R., Speech Recognition by Machine: a Review, *Proc. IEEE*, **64**, 501–531 (1976).

[16] Smith, F. J., and Linggard, R. J., Information Retrieval by Voice Input and Output, *Lect. Notes Comp. Sci.*, **146**, 275–288 (1983).

[17] Erman, L. D., Hayes-Roth, F., Lesser, V. R., and Reddy, D. R., The Hearsay-II Speech-understanding System: Integrating Knowledge to Resolve Uncertainty, *Comp. Surveys*, **12**, 213–253 (1980).

[18] Foley, J. D., and van Dam, A., *Fundamentals of Interactive Computer Graphics*, Addison-Wesley, New York (1982).

[20] Gund, P., Three-dimensional Pharmacophoric Pattern Searching, *Prog. Mol. Subcell. Biol.*, **5**, 117–143 (1977).

[21] Gund, P., Andose, J. D., Rhodes, J. B., and Smith, G. M., Three-dimensional Molecular Modelling and Drug Design, *Science*, **208**, 1425–1431 (1980).

[22] Langridge, R., Ferrin, T. E., Kuntz, I. D., and Connolly, M. L., Real-time Color Graphics in Studies of Molecular Interactions, *Science*, **211**, 661–666 (1981).

[23] Sammet, J. E., An Overview of High-level Languages, *Advances in Computers*, **20**, 199–259 (1981).

[24] Treleaven, P. C., and Lima, I. G., Japan's Fifth-generation Computer Systems, *Computer*, **15**, 79–88 (1982).

[25] Jensen, K., and Wirth, N., *PASCAL User Manual and Report*, Springer Verlag, New York (1975).

[26] Barnes, J. G. P., An Overview of ADA, *Software Pract. Exp.*, **10**, 851–887 (1980).
[27] Hoare, C. A. R., The Emperor's Old Clothes, *Comm. ACM*, **24**, 75–83 (1981).
[28] Kernigan, B. W., and Morgan, S. P., The UNIX Operating System: a Model for Software Design, *Science*, **215**, 779–783 (1982).
[29] Brooks, F. P., *The Mythical Man-Month*, Addison-Wesley, Reading, Mass. (1975).
[30] Deen, S. M., *Fundamentals of Data Base Systems*, Macmillan, London (1977).
[31] Date, C. J., *An Introduction to Data Base Systems*, Addison-Wesley, Reading, Mass. (1981).
[32] Rowley, J. E., *Mechanised In-house Information Systems*, Bingley, London (1979).
[33] Ashford, J. H., Information Management Packages on Minicomputers, *J. Inform. Sci.*, **2**, 23–28 (1980).
[34] Tagg, R. M., Bibliographic and Commercial Databases – Contrasting Approaches to Data Management with Special Reference to DBMS, *Program*, **16**, 191–199 (1982).
[35] Macleod, I. A., and Crawford, R. G., Document Retrieval as a Database Application, *Inform. Technol. Res. Develop.*, **2**, 43–60 (1983).
[36] McTaggart, J. A., and Radcliffe, J., The Construction and Organization of Computerized Toxicology Databases, *Inform. Sci.*, **11**, 101–111 (1977).
[37] Ravenscroft, T., and Smith, D. E., The Development of CLINDATA, a Clinical Trial Data Management System, *J. Inform. Sci.*, **3**, 129–136 (1981).
[38] Raitt, D. I., Recent Developments in Telecommunications and their Impact on Information Services, *Aslib Proc.*, **34**, 54–76 (1982).
[39] Cawkell, A. E., Information Technology and Communications, *Ann. Rev. Inform. Sci. Technol.*, **15**, 37–65 (1980).
[40] Casey, M., Packet Switched Data Networks: an International Review, *Inform. Technol. Res. Develop.*, **1**, 217–244 (1982).
[41] Clarke, A. C., Extra Terrestrial Relays: Can Rocket Stations Give World-wide Radio Coverage?, *Wireless World*, **51**, 69–75 (1945).
[42] Edelson, B. I., Global Satellite Communications, *Sci. Am.*, **236**, 58–69 (1977).
[43] Kelly, P. T. F., The EURONET-DIANE Project, *Int. Forum Inform. Doc.*, **7**, 22–27 (1982).
[44] Clarke, D. D., Pogran, K. T., and Reed, D. P., An Introduction to Local Area Networks, *Proc. IEEE*, **66**, 1497–1517 (1978).
[45] Shock, J. F., Dalal, Y. K., Redell, D. D., and Crane, R. C., Evolution of the Ethernet Local Computer Network, *Computer*, **15**, 10–27 (1982).
[46] Collinson, R. P. A., The Cambridge Ring and UNIX, *Software Pract. Exp.*, **12**, 583–594 (1982).

11

Trends in the communication of chemical information

The previous chapters have described the many new techniques and technologies that are emerging for the storage, retrieval and communication of chemical information. Almost without exception, these developments have arisen from the increasing power, flexibility, and availability of computer systems, and there is little reason to doubt that the current rapid growths in processing speeds, storage capacities etc. will continue within the near future. Therefore the power of present minicomputers is likely to be available in desktop workstations within two to three years, while current supercomputer characteristics will increasingly be incorporated into mainframes and minicomputers; in particular, there will be a move towards the incorporation of an increasing degree of parallelism in computer systems, this permitting the use of molecular orbital calculations and real-time colour graphics *inter alia* on a routine basis in drug design programmes. The developments in data storage devices, in particular of erasable digital optical discs, offer the prospect of virtually limitless online storage at much lower costs than those associated with magnetic disc systems, and provide the means for the storage and searching of the full texts of documents, rather than the title- and abstract-based files that have been available to date.

Given the rapidly changing technological picture, it is impossible to predict accurately the future development of chemical information systems, but three trends of potential importance may be noted. The first of these is a movement back to the chemist as the end-user of the data available in machine-readable stores, and this trend is certain to gather pace as the major online information providers start to promote the use of their systems by the chemist himself. The 1960s and 1970s saw the chemist becoming divorced from the literature owing to the enormous volume of information produced, to the diffusion of this information among many new specializations, and to the clumsy searching techniques which were available. This isolation arose from the emergence of information specialists, who were trained in the use of indexing languages, in the design of search strategies, and with experience of the searching mechanisms provided by the various database vendors. These specialists acted as intermediaries between the chemist and the computer systems available to search the scientific literature. The gradual introduction of more user-friendly hardware and software interfaces has resulted in the greater acceptability of these systems to the chemist. For example, terminals are available now which enable automatic dial-up and connection to remote hosts, and which provide assistance in the formulation of queries and the local reformatting and manipulation of retrieved data. The acceptance of poorly drawn structure diagrams and trivial chemical names as search parameters, and the ability to browse through abstracts and full-text files, all promote the direct involvement of the chemist in his search for published information. Once the chemist again becomes responsible for identifying and satisfying his own information needs, the role of the information specialist should become that of promoting the use of information, of advising on the existence and applications of information systems, and of creating and controlling new databases and databanks.

The second major development is likely to be in the nature of the major providers of information. The online information industry was born out of the spare computing capacity of large business and industrial organizations, who undertook to provide the hardware and software facilities for mounting and searching databases generated by publishing companies. The industry has evolved to the stage at which the producer of the information may now seek to be involved not only in the production of the database but also in its distribution, in both printed and electronic forms. This evolutionary trend has serious implications in chemistry since it increases the dominance of Chemical Abstracts Service (CAS) in a market in which it is already the major force. In the 1970s, CAS information was made available by a range of vendors, so allowing considerable scope for imagination in the repackaging of the data and in bringing the disciplines of the competitive market to bear. As CAS moves to regain control of the marketing and distribution of its data, primarily in order to secure from online services the revenue which is being lost through declining sales of its printed products, there will exist the possibility of an undesirable monopoly in chemical information. Although caused by financial necessity, this potential monopoly is viewed with some suspicion by many information scientists; it should be noted, however, that many other major services, such as those produced by Beilstein, the National Library of Medicine, or the Institute for Scientific Information, are independent of CAS. Additionally, there is an increasing number of databases that do not derive from a printed literature source, and that have been designed only for use in machine-readable form.

The third and, perhaps, most important development is in the increasing integration of services and data files that were previously used only in isolation. An example is the ability to carry out highly specific structure and subject retrieval by performing a substructure search on a structure file, and then transferring automatically the Registry Numbers retrieved to a subsequent search of a textual database, so as to retrieve the bibliographical details of documents which refer to the synthesis, properties or uses of those compounds. Other examples of integration include the sophisticated searches that can be carried out in the National Institutes of Health/Environmental Protection Agency complex of databases and databanks, the use of machine-readable files of reactions in combination with computer-aided synthesis design programmes, the use of structural and biological information in conjunction for studies of quantitative structure–activity relationships, the searching of unified files of the journal and patent literature using current substructure search techniques, and the linkage of in-house systems with publicly available services.

The growing number of information files available for public or in-house searching will bring many novel and creative uses of databases and the information that they contain. Apart from legal implications, such as the law of copyright, this growth is likely to be limited only by two factors. The first of these is the problems associated with the interconnection of different, and often incompatible, systems; such difficulties can, in principle, be overcome by the general acceptance of standard interfaces and protocols. The second factor which determines the rate of uptake of the new technologies will be the financial situation of those organizations that need to use them. It is already the case that the use of online bibliographic databases by academic and public library services lags far behind that of industrial users, largely because of the lack of adequate finance for these services. This disturbing tendency can only increase with the introduction of the more sophisticated, but expensive, structure databases and databanks. In spite of increased costs in staffing and in the obtaining of the source material, the financial outlook for database producers, conversely, is rather more favourable since the rapidly decreasing cost of computing means that even small organizations can now apply automated production methods, with the result that the basic primary and secondary publications can be produced more efficiently, and a wide range of additional services can be offered at minimal extra cost. In the longer term, the use of electronic journals and automated document delivery systems may herald the end of printed media as the primary means of communication, with hard copies being produced only when required, for example in response to searches of the electronic archives. However, such changes can take place only when the database producers are assured that their costs can be recouped from computerized searching alone, and this is unlikely to be the case for many years yet; however, as noted above in the case of CAS, database producers are already experiencing a noticeable diminution in revenue from the printed versions of their databases.

In conclusion, developments in computing, telecommunications, structure and data handling techniques, and molecular graphics provide a real opportunity to create the 'ideal' chemical information system. What seems to be lacking at present is the mechanism, or the will, to achieve the cooperation required to attain this objective. Information is now a business which, like all other businesses, is competitive in nature, and offers the prospect of great rewards, both financial and otherwise. Whilst the scientific community has benefited from the diversity of services resulting from this competition, there is now a need to ensure a much greater degree of integration than heretofore. Only the users of chemical information systems can ensure that this unification of techniques and services takes place; if it does, the chemical information nirvana *may* become a reality.

Appendix 1

Glossary of acronyms, trade and product names

ACMF	Augmented Connectivity Molecular Formula (CAS)
ACS	American Chemical Society
ACSP	Advisory Council on Scientific Policy (UK)
ACT	Computer firm
ADAPT	Automated Data Analysis using Pattern Recognition Techniques
AI	Artificial Intelligence
ALCHEM	Associative Language for Chemistry (SECS)
ALWIN	Algorithmic Line Notation based on Wiswesser
AQUIRE	Aquatic Information Retrieval databank (CIS)
ARDIC	Association pour la Recherche et le Développement en Informatique Chimique
ARTHUR	Data Analysis and Pattern Recognition Component (CIS)
ASLIB	The Association of Information Management (formerly the Association of Special Libraries and Information Bureaux)
AWRE	Atomic Weapons Research Establishment (UK)
BASIC	Basle Information Centre for Chemistry
BASIC	Beginners' All-Purpose Symbolic Instruction Code (a computer programming language)
BLAISE	British Library Automated Information Service
BLEND	Birmingham and Loughborough Electronic Network Development (UK)
BLLD	British Library Lending Division
BRS	Bibliographic Retrieval Service (US)
BT	British Telecom
BUCOP	British Union Catalogue of Periodicals
CA	*Chemical Abstracts*
CAB	Commonwealth Agricultural Bureaux (UK)
CAC&IC	*Current Abstracts of Chemistry and Index Chemicus* (ISI)
CAIRS	Computer Assisted Information Retrieval System
CAOCI	Commercially Available Organic Chemicals Index (now FCD)
CARI	Chemical Abstracts Review Index

CAS	Chemical Abstracts Service
CASIA	Chemical Abstracts Subject Index Alert
CASP	Computer Aided Synthesis Program
CASSI	Chemical Abstracts Service Source Index
CCRIS	Chemical Carcinogenesis Research Information System (CIS)
CDST	Centre de Documentation Scientifique et Technique (CNRS)
CEA	*Chemical Engineering Abstracts*
CESARS	Chemical Evaluation Search and Retrieval System (CIS)
CFR	Code of Federal Regulations (US)
CHEMICS	Combined Handling of Elucidation Methods for Interpretable Chemical Structures (CIS)
CHEMLAB	Chemical Modelling Laboratory (CIS)
CHEMLAW	Code of Federal Regulations: Chemical Regulations (CIS)
CHMTRN	Chemistry Translator (LHASA)
CIMI	Chemical Information Management Inc.
CIS	Chemical Information System compiled jointly by the US Government Agencies NIH and EPA
CNA	Chemical Notation Association
CNIC	Centre National de l'Information Chimique
CNMR	Carbon-13 Nuclear Magnetic Resonance Spectral Search System (CIS)
CNRF	Centre National de la Recherche Scientifique, Paris
COBOL	Common Business Oriented Language (a computer programming language)
CODATA	Committee on Data for Science and Technology
CORA	Semiautomatic Coding System for Markush Structures (Roussel-Uclaf)
COUSIN	Compound Information System (Upjohn Co.)
CPI	Central Patents Index (Derwent Publications Ltd)
CROSSBOW	Computerized Retrieval of Organic Structures Based on Wiswesser (Fraser Williams (Scientific Systems) Ltd)
CRT	Cathode Ray Tube
CRYST	X-ray Crystallographic Search System (CIS)
CSA	Chemical Structure Association
CSD	Cambridge Structural Database (UK)
CSI	*Chemical Substance Index* (CAS)
CSI	*Chemical Substructure Index* (ISI)
CSSR	Crystal Structure Search and Retrieval (SERC implementation of CSD)
CTCP	Clinical Toxicology of Commercial Products (CIS)
DAP	Distributed Array Processor (ICL)
DARC	Description, Acquisition, Retrieval and Correlation (Télésystèmes-Questel)

DARING	WLN-Connection Table Conversion Program (SERC)
DBMS	Database Management System
DEC	Digital Equipment Corporation
DIALOG	Online Information Retrieval System (Lockheed Inc.)
DIMDI	Deutsches Institut für Medizinische Dokumentation und Information
DOC5	Dictionary of Organic Compounds (fifth edition)
ECDIN	Environmental Chemicals Data and Information Network
ECOIN	European Core Inventory
EEC	European Economic Community
EINECS	European Inventory of Existing Chemical Substances
EM	Ensemble of Molecules
EPA	Environmental Protection Agency (US Government)
EPC	European Patent Convention
EPO	European Patent Office
EROS	Elaboration of Reactions for Organic Synthesis
ESA-IRS	European Space Agency – Information Retrieval Service
EURECAS	CA Registry Search File (Télésystèmes-DARC)
EUSIDIC	European Scientific Information Dissemination Centres
FCD	Fine Chemicals Directory (Fraser Williams (Scientific Systems) Ltd)
FDA	Federal Drug Administration (US Government)
FORTRAN	Formula Translator (a computer programming language)
FRSS	Federal Register Search System (CIS)
FTX	Free Text
GENSAL	Generic Structure Language (University of Sheffield)
GOLEM	Online Structure Input System (IDC)
GRAI	Government Reports Announcement and Index (US)
GREMAS	Generic Retrieval by Magnetic Tape Search (IDC)
GRSC	Graduateship of the Royal Society of Chemistry (UK)
HSE	Health and Safety Executive (UK)
IBM	International Business Machines
IC	*Index Chemicus* (ISI)
ICI	Imperial Chemical Industries
ICL	International Computers Limited
ICRS	Index Chemicus Registry System (ISI)
ICSU	International Council of Scientific Unions
IDC	Internationale Dokumentationsgesellschaft für Chemie
IEEE	Institute of Electrical and Electronic Engineers
IFI	Information for Industry/Plenum Corporation
IIB	Institut International des Brevets
INKA	Information System Karlsruhe Online Service (FDR)
INTELSAT	International Telecommunications Satellite Organization

IPSS	International Packet Switched Service
IRCS	International Research Communications System (Medical Science Journal)
IRSS	Infra-red Search System (CIS)
ISI	Institute for Scientific Information
IUPAC	International Union of Pure and Applied Chemistry
KWIC	Keyword in Context
LAN	Local Area Network
LAYOUT	2-Dimensional Structure Display Package (MDL)
LHASA	Logic and Heuristics Applied to Synthetic Analysis
LSI	Large Scale Integration
MACCS	Molecular Access System (MDL)
MALIMET	Master List of Medical Terms (*Excerpta Medica* thesaurus)
MCS	Maximal Common Substructure
MDL	Molecular Design Limited Inc.
MLAB	Mathematical Modelling Laboratory (CIS)
MR	Molar Refractivity
MS	Mass Spectrometry
MSD	Mass Spectrometry Data Bank (UK)
MSDC	Mass Spectrometry Data Centre (UK)
MSSS	Mass Spectral Search System (CIS)
NBS	National Bureau of Standards (US)
NIH	National Institute of Health (US Government)
NIH-EPA CIS	see CIS
NIOSH	National Institute of Occupational Safety and Health (US)
NLM	National Library of Medicine (US)
NMR	Nuclear Magnetic Resonance
NPL	National Physical Laboratory (UK)
NTIS	National Technical Information Service (US)
NUCSEQ	Nucleotide Sequence Search System (CIS)
OCCI	Organic Chemistry Citation Index (ISI)
OCR	Optical Character Recognition
OCSS	Organic Chemical Simulation of Synthesis
OLSIS	Online Structure Input System (CAS)
OHMTADS	Oil and Hazardous Materials – Technical Assistance Data System (CIS)
ORBIT	Online Information Retrieval System (SDC)
OSTI	Office of Scientific and Technical Information (now British Library R&D department)
PCT	Patent Co-operation Treaty
PDR	Pharma Dokumentationsring
PIRA	Paper Industries Research Association
PR	Pattern Recognition

PSS	Packet Switched Service (UK)
PSTN	Public Switched Telephone Network
QI	Quality Index
RADIICAL	Retrieval and Automatic Dissemination of Information from *Index Chemicus* and Line Notations (ISI)
RAPRA	Rubber and Plastics Research Association
RIN	Ring Identifier Number (CAS)
RN	Registry Number (CAS)
RNF	Registry Nomenclature File (CAS)
RS	Royal Society (UK)
RSC	Royal Society of Chemistry (UK)
RTECS	Registry of Toxic Effects of Chemical Substances
SANSS	Structure and Nomenclature Search System (CIS)
SAR	Structure Activity Relationship
SCI	Science Citation Index (ISI)
SDC	System Development Corporation (US)
SDI	Selective Dissemination of Information
SECS	Simulation and Evaluation of Chemical Synthesis
SEMA	Stereochemically Extended Morgan Algorithm (MDL)
SERC	Science and Engineering Research Council (UK) formerly SRC
SIMD	Single Instruction Stream, Multiple Data Stream
SISD	Single Instruction Stream, Single Data Stream
SKF	Smith Kline & French Co. Ltd.
SOCRATES	Sandwich Online Chemical Retrieval and Topological Evaluation System (Pfizer UK)
SPHERE	Scientific Parameters in Health and the Environment Retrieval and Estimation (CIS)
SRC	Science Research Council (now SERC)
SRIM	Selected Research in Microfiche
SYNCHEM	Synthetic Chemistry program
TELENET	Telecommunications Network
THERMO	Thermodynamics (CIS)
TOXLINE	Toxicology Database Available Online from the NLM
TSCA	Toxic Substances Control Act (US)
TSCAPP	Toxic Substances Control Act Plant and Production data (CIS)
TYMNET	Time Sharing Network
UCRR	Unique Chemical Registry Record (CAS)
USAN	United States Adopted Name
UVCB	Substances of Unknown or Variable Composition, Complex Reaction products and Biological Materials
VDU	Visual Display Unit
VLSI	Very Large Scale Integration
WLN	Wiswesser Line-formula Notation

WPI	*World Patents Index* (Derwent Publications Ltd)
WPIL	*World Patents Index Latest* (Derwent Publications Ltd)
WRAIR	Walter Reed Army Institute of Research
WSCA	World Surface Coatings Abstracts (Pergamon-InfoLine database)
XTAL	NBS Crystal Data Identifications File (CIS)
ZLCA	Zinc, Lead and Cadmium Abstracts (Pergamon-InfoLine database)

Appendix 2

Organization addresses

The addresses given below are of organizations which may be contacted for further information on systems and services described in this book.

ASLIB	Aslib, 3 Belgrave Square, London SW1X 8PL.
BASIC	Basle Information Centre for Chemistry, CH-4002 Basle, Switzerland.
Beilstein	The Beilstein Institute, Varentrappstrasse 40-42, D-6000, Frankfurt/M. 90, Federal Republic of Germany.
BLAISE	The British Library Automated Information Services, British Library Bibliographic Services Division, 7-9 Rathbone Street, London W1P 2AL.
BLLD	British Library Lending Division, Boston Spa, Wetherby, W. Yorks LS23 7BQ, UK.
BRS	Bibliographic Retrieval Services, Inc. 1200 Route 7, Latham, New York 12110, USA.
CAS	Chemical Abstracts Service, P.O. Box 3012, Columbus, Ohio 43210, USA.
CIMI	Chemical Information Management Inc., P.O. Box 2740, Cherry Hill, New Jersey 08034, USA.
CNIC	Centre National de l'Information Chimique, 28 ter, rue Saint-Dominique, 75007 Paris, France.
Data-Star	Data-Star, Willoughby Road, Bracknell, Berkshire, RG12 4DW, UK.
Derwent	Derwent Publications Ltd, Rochdale House, 128 Theobalds Road, London WC1X 8RP.
DIALOG	Dialog Information Retrieval Services, Inc., 3460 Hillview Avenue, Palo Alto, California 94304, USA, *or* Dialog Information Services, Besselsleigh Road, Abingdon, Oxford OX13 6LG, UK.
DIMDI	DIMDI, Weisshausstrasse 27, Postfach 42 05 80, D-5000 Köln 41, Federal Republic of Germany.

ECDIN	Commission of the European Communities, Joint Research Centre, Ispra Establishment, 21020 Ispra (Varese), Italy.
ESA-IRS	ESA Information Retrieval Service, via Galileo Galilei, 00044 Frascati, Italy.
EUSIDIC	EUSIDIC, P.O. Box 429, London W4 1UJ.
Fraser Williams (Scientific Systems)	Fraser Williams (Scientific Systems) Ltd, London House, London Road South, Poynton, Cheshire SK12 1YP, UK.
IDC	IDC, Hamburger Allee 26-28, 6000 Frankfurt am Main 90, Federal Republic of Germany.
Inka	Inka, c/o Fachinformationszentrum Energie, Physik, Mathematik GmbH, D-7514 Eggenstein-Leopoldshafen 2, Federal Republic of Germany.
ISI	The Institute for Scientific Information, 3501 Market Street, University City Science Center, Philadelphia, PA 19104, USA.
MDL	Molecular Design Ltd, 1122 B Street, Hayward, California 94541, USA.
NIH-EPA CIS	CIS User Support, Computer Sciences Corp., P.O. Box 2227, Falls Church, Virginia 22042, USA.
NLM	National Library of Medicine, Bethesda, Maryland 20209, USA.
Pergamon-InfoLine	Pergamon-InfoLine Ltd, 12 Vandy Street, London EC2A 2DE.
RSC	Royal Society of Chemistry, The University, Nottingham NG7 2RD, UK.
SDC	System Development Corporation, 2500 Colorado Avenue, Santa Monica CA 90406, USA.
STN	STN-Columbus, c/o Chemical Abstracts Service, 2540 Olentangy River Road, P.O. Box 2228, Columbus, Ohio 43202, USA, *or* STN-Karlsruhe, c/o Fachinformationszentrum Energie, Physik, Mathematik GmbH, D-7500 Karlsruhe 1, Postfach 2465, Federal Republic of Germany.
Télésystèmes-Questel	Télésystèmes-Questel, Tour Gamma B, 193-197 rue de Bercy, 75582 Paris Cedex 12, France.
Upjohn Co.	Upjohn Co., Kalamazoo, Michigan 49001, USA.

Index

Q

R

T